Our Oldest Companions

Our Oldest Companions

THE STORY OF THE FIRST DOGS

Pat Shipman

The Belknap Press of Harvard University Press

CAMBRIDGE, MASSACHUSETTS

LONDON, ENGLAND

2021

First printing

LIBRARY OF CONGRESS CATALOGING-IN-PUBLICATION DATA

Names: Shipman, Pat, 1949– author.
Title: Our oldest companions : the story of the first
dogs / Pat Shipman.
Description: Cambridge, Massachusetts : The Belknap
Press of Harvard University Press, 2021. | Includes
bibliographical references and index.
Identifiers: LCCN 2021012213 |
ISBN 9780674971936 (cloth)
Subjects: LCSH: Dogs—Evolution. | Dogs—Effect of
human beings on. | Coevolution. | Human-animal
relationships. | Human evolution. | Paleontology.
Classification: LCC QL737.C22 S5375 2021 |
DDC 599.77/2—dc23
LC record available at https://lccn.loc.gov/2021012213

*To Pearl, the Wonder Dog, who taught me more
about dogs than anyone before or since*

Contents

Contents

Preface

We humans like narratives. We are comfortable with a beginning, a middle, and an end. Stories make sense to us. I wonder if the human brain is hardwired to make stories out of important pieces of information.

When we look back into the past to see how we became who we are today, we usually tell the story backward, as if it started today rather than long ago. If, for example, we were interested in the Nile River, we might look for its origin by tracing it backward from where it empties into the ocean. We would follow the main flow, ignoring or not even noticing the many rivulets and streams that branch off from it as well as those that join it somewhere and add to its flow. We search for beginnings, almost without thinking about it, and too often ignore what look like minor deviations. This tendency sometimes distorts the actual meaning of events. Looking backward, we are blind to the myriad alternatives that didn't occur—or that did, but that didn't work well enough to produce a lasting change. The long chain of serendipitous accidents in our history is pared down when we look backward. The failures, the die-offs, the influence of random occurrences nearly all disappear in hindsight. Arriving at the present situation begins to seem preordained, scripted in advance, *intended.* Nothing could be further from the truth when we are talking about evolution and the staggeringly influential changes evolution brought to human lives.

In this book, I am going to try to tell the story of one of the greatest discoveries in human history—but also to tell it the right way around. My tale is about the way in which we learned how to evolve without literally evolving and how to adapt without changing our physical characteristics. I want to tell the story of how people learned to work with other species so that we could borrow their exceptional abilities without having to evolve them ourselves. This is the story of the first dog and its humans.

Our partnership with other species is often called *domestication.* I do not like that word. First, it is used too broadly, to apply to both plants and animals, which surely have widely different experiences of domestication attempts. Second, it is also used too narrowly, to apply only to species in which humans have tightly controlled the animals' reproduction. Both are inaccurate.

There is also a widespread assumption that domestication benefited humans but not the other species that were our partners in this endeavor—and that belief is wrong, too. A few animals have been domesticated, but many more have not. Some animals have been targeted for domestication but have (consciously or not) simply refused. You might think zebras should be capable of being domesticated, since horses and donkeys are, right? Nope. You may see photos from the late 1800s or early 1900s, taken by colonial settlers in Africa, showing carriages with zebras in the traces or even zebras with saddles on. But when you read the captions closely, you realize that those zebras were certainly not domesticated. They regularly kicked carriages or carts to pieces and refused to cooperate under a saddle. They bit and were hard to handle. Some zookeepers claim that zebras are the most dangerous and aggressive animals in the zoo, so reluctant are they to accept human authority.[1]

The reality is that the mammals that have been domesticated have cooperated, have *chosen* to associate with humans, and have actively participated in forging a new life that involves intimacy with humans. For some species, their adaptive niche incorporates the human niche or the anthropogenic (human-created) environment. The species whose environment or niche overlaps heavily with ours are those that we consider domesticated. What these animals evolved to do is to live with humans and form bonds with humans. Of all the domestic animals in the world, the dog is arguably the most thoroughly domesticated. It was certainly the first to be domesticated.[2]

Some people have supposed that past humans captured a baby animal, tamed it, brought it up, selected a mate for it, brought up those offspring (keeping the nice, friendly babies and killing or abandoning the others) and so on until—hey presto!—wolves became dogs, the mighty aurochs became cattle, the agile mountain goats became domestic goats. Others suppose that animals somehow domesticated themselves once they discovered that humans sometimes had leftover food. These are fairy tales. The true story—the real story—is much more nuanced.

Part of the task I set myself in writing this book was to figure out why we have so often misunderstood the full story. This book describes how we became integrated with the other animals of this earth in a new and different way. That integration—that twisting, surprising, and sometimes dumb-founding set of events—made all the difference in our human lives. That integration gave humans access to evolutionary shortcuts and a much wider range of abilities than we our-selves possess.

There are still dark and shadowy corners in this narrative, places where we have not yet found enough evidence to see exactly what happened, and especially why. But there are new threads and themes and influences in my tale that I think—I hope—have led me closer to the truth than ever before.

A last word: Parts of this book concern dingoes, and Greater Australia, and Australian Indigenous peoples. In my attempt to explore and re-create the actions of people from the deep human past, I have done my best to accurately represent the remarkable Indigenous people of that continent and their traditions.

Our Oldest Companions

ONE

Before Dogs

FOR MILLENNIA, humans' most constant companions have been dogs. To talk about the history of the dog is to talk about how this ubiquitous, highly variable, and oft-beloved creature both came to be and came to be with us, almost literally one of us. We call dogs our best friends, and in many senses, that is true. We buy them food, toys, clothing, beds, and much more, as if they were children. We build them houses or special nooks or shelters within our own houses. It was during the process of entering into a pact with the future dog that we learned how to domesticate and work collaboratively with another species, but there was no intentionality in the process. Nobody set out to "make" a dog. The dog that we know so well, the species that lives with us virtually everywhere in the world, warms our beds, plays with our children, herds our sheep, protects our dwellings, and helps us locate and kill prey was not consciously sought or created. If it had been, nobody in their right mind would have started with an ancient wolf.

As Bronwen Dickey has shown so clearly in her writing about pit bulls, dogs today are often proxies for humans. In fact, as she argues, sometimes when people express vehement fear and dislike of a particular breed of dog, as often occurs with pit bulls, they are really speaking about the type of humans that they believe most often own those dogs. It is much more acceptable to speak against the dogs than against the people. So, when we talk about dogs and the origin and evolution of dogs, we are also talking about people. In some cultures, dogs and people are virtually interchangeable. This is perhaps the clearest sign that doggy creatures are truly domesticated.[1]

Archaeologists and paleontologists also make use of this principle. Sometimes it is easier to find out about human migrations and settlements by studying the dogs and not the people. Elizabeth Matisoo-Smith of the University of Otago in New Zealand pioneered this so-called commensal approach for genetic studies. (*Commensal* is an ecological term for a type of relationship between two species that live together in which one benefits from the presence of the other but the other remains less obviously affected by the relationship.) Matisoo-Smith observed that humans often travel with and transport other species, particularly in the Pacific islands and continents, where much travel occurs by boat. Pigs, chickens, rats, and dogs, which are widespread in Polynesia, can't have traveled by themselves from one island to the next. She realized that looking at the genetics of species that live with and are transported by humans could yield important information about migration pathways. Thus commensal studies can confirm or even substitute for genetic studies of humans, living or ancient.[2]

Fairly often, establishing a commensal relationship with another species seems to have been the beginning of the domestication process. I would argue, though, that mutualism—a relationship in which both species benefit and hold a shared set of values—is a better characterization of the basis of domestication. Mutualism most probably applies to dogs and their relationships to humans, in any case.

It is vital to remember that, when the first dog appeared, no animal had ever entered into such a collaboration with humans, and no human, I feel sure, had ever imagined such a state. No one imagined a wolf dozing on a specially bought bed in front of the fire. Yet it happened. The improbable does happen, more often than we appreciate. The road to living with dogs must have been unfamiliar and strange.

In the United States alone, more than 73 million dogs live with humans; worldwide, there are an estimated 900 million dogs. They live in human homes or in the wild on every continent except Antarctica, from which the last dog was deliberately removed in 1994. In the Western world, many dogs are thought of as family members; others are working dogs bred and trained to specific tasks, coworkers rather than kin. Some are both. Some dogs, in some cultures, are food items. This is true now and has been in the past.

But let us not focus on the surprising end result, an animal known for its extraordinary variability in shape, size, color, and behavior. Let's ask what happened then, in the deep past, when there was no such thing as a dog.

In the beginning, there was the gray wolf, *Canis lupus*. One of the most thoughtful and knowledgeable scholars studying the genetics and evolution of canids (the zoological family that includes wolves, dogs, dingoes, foxes, jackals, and other similar species) is Robert (Bob) K. Wayne of the University of California, Los Angeles, whom I have known since he first started studying canid genetics in graduate school. After decades of work on canids, he remarked, "*Dogs are gray wolves,* despite their diversity in size and proportion." And yet, of course, we all know that dogs are not gray wolves. Therein lies the paradox. The genetic separation of domestic dogs and gray wolves is slight: "they differ by at most 0.2% of mtDNA sequence," according to Bob and his collaborators.[3] (Mitochondrial DNA, or mtDNA, is one of two types of DNA in most organisms; it is passed through the maternal line only and is found in cell organelles called mitochondria. The other is nuclear DNA, which combines genes from both parents and is contained in the nucleus.) Yet the behavioral separation is substantial. The dog is our most faithful, numerous, and widespread companion. The wolf is among the most feared and hated of predators, one very often targeted for destruction by humans.

There are five potential settings in which the evolution of the first dog might have occurred: Africa, Europe, Asia, Australia, and the Americas. Australia can be largely eliminated because it has no history of ancient or fossil canids that could have evolved into dogs, although it now has the dingo (only debatably a dog). Africa has the newly recognized golden wolf—long mistaken for a jackal—and the Ethiopian wolf which, confusingly, is actually a jackal. There are thought to be small, isolated populations of gray wolves in northern Africa, but gray wolves may never have been common on the continent. But Africa is also the place where modern humans first evolved, and the domestication of any species requires the presence of humans. The Americas might be the locus for dog evolution—or for an independent domestication of dogs from gray wolves. Though this is at least theoretically possible, dogs would have had to travel with humans from the Americas back into Eurasia. A massive expansion of humans from the Americas back into the Old World is not generally regarded as likely or supported by significant evidence. This leaves Europe and Asia as the main contenders for the birthplace of the first dog.[4]

It has long been assumed that the first dog arose in Europe. The first main reason this assumption was made was that, during the era in which this tradition became established, European scholars were ignorant about and largely dismissive of Asia, Africa, and Australia as places of innovation. Second, because most early paleontologists were Europeans, there was a tendency for evidence to be sought in Europe. However, as information from around the globe has been collected and more rigorously evaluated in recent decades, the support for either a European origin or an Asian one—or both—has grown.

Where and when in the past were both wolves and modern humans present—the essential ingredients—so that a domestic dog could possibly arise?

The gray wolf evolved in Eurasia about 800,000 years ago, well before our ancient cousins, Neanderthals, evolved there about 400,000 years ago. Both the European wolves and Neanderthals were predators, members of the same ecological role, or *guild,* as ecologists term it. This implies a certain similarity of outlook, a resemblance between their lifestyles, a commonality of habits and resources. Neanderthals and European wolves preyed upon the same animals, and they often used the same caves for sleeping, raising young, and keeping warm.

Before the first modern humans arrived in Europe, the region was inhabited by a diverse set of predators. There were enormous cave lions and powerful cave hyenas, big leopards, cave bears, large-bodied wolves, and packs of sturdy doglike animals most similar to today's dholes—the stocky wild canids of Asia. And there were of course Neanderthals, who had evolved and been part of the European ecosystem for about 400,000 years. Like them, and like our early modern ancestors, all of these fearsome predators focused on hunting medium- to large-sized herbivores. That means there was stiff competition for prey, both live and dead. Stealing a carcass is a great way to get a meal if you can drive off the other predator that made the kill. If not, you don't get anything to eat and you may be fatally injured. But more than food was important—so, too, were places to find refuge, to drink, to sleep, to raise young, to hibernate, to sit out bad weather.

To what extent did this competition between Neanderthals and other European predators influence their behavior? A great deal, apparently. A recent, detailed study of the use of cave sites by Neanderthals, cave bears, cave hyenas, cave lions, leopards, and wolves, led by French paleontologist Camille Daujeard of the Institut de Paléontologie Humaine, reveals some fascinating ways in which the European Ice Age predators divided up these resources and lessened competition.[5] The team compared cave use in the Central Massif of France with that in the Meuse Basin of Belgium, a thousand kilometers away, from about 108,000 years to 30,000 years ago. This time span includes the period about 50,000 to 40,000 years ago, during which modern humans first arrived in Europe after they expanded out of Africa. Caves were apparently one of the valuable resources used by several carnivores, including the archaic humans, or hominins, that are known as Neanderthals.

Daujeard and her colleagues looked at the shapes of caves, at fossilized animal bones, at stones modified by humans, and at damage or textures on the ancient bones that attested to their history, such as carnivore tooth marks, cut marks, or burn marks. When team members finished their statistical analysis, they found that the shape of the caves was an important determinant of which species used them.

Carnivores were the predominant occupier of caves with small chambers. If the chambers were deep, they were ideal places for cave hyenas to den and raise their cubs and for cave bears to hibernate. They left behind the intensely gnawed bones of their prey and sometimes the bones of their offspring. Small groups of Neanderthals, who left relatively clean bones of prey and some stone artifacts, used this type of cave much more rarely. The team suggests these caves were used by Neanderthals as seasonal hunting camps, not for long-term occupation.

Wolves preferred narrow caves with high ceilings and steep accesses, as did bears and the larger felids. The fossil remains left in the caves showed that wolves and cave bears often used caves of this shape for hibernating, denning, and sheltering. Occasionally small herbivores made the risky choice of sheltering inside the caves in bad weather, increasing the danger of death by literally walking into a predator's den. Neanderthals did not commonly use caves with this shape, although groups of limited size sometimes used the caves as bivouacs.

Caves with enormous entrances or rock shelters—natural overhangs—were the preferred choice of larger groups of Neanderthals, and they stayed longer in these caves. The remains of carnivores are relatively scarce in these caves, and the traces of carnivore modification, such as chewing marks on bones, are rare. Either Neanderthals kept the carnivores out of such places or the carnivores simply favored caves of other shapes.

This insightful analysis by Daujeard and her colleagues proved statistically that particular predators preferred caves of particular shapes. This work enables us to see how the varied predatory species had adapted to the presence of the other species and their needs—they competed but lessened the competition by using the resources differently. Neanderthals lived in fairly small groups that moved around the country-side and focused on using known, local resources (prey, stone, water, and caves).

Neanderthals had evolved in Europe from other archaic humans, so they had had a long time to adjust to the habitat, local topography, and local ecosystem. During their evolution, they worked out how to avoid the habitats the other preda-tors favored and how not to become someone else's dinner while hunting for their own. The fossil and archaeological evi-dence indicates that Neanderthals were probably never very numerous, and their genetic diversity is unusually low. In short, they were inbred. Nonetheless, they were reasonably successful, surviving despite the many predators that shared their region. However, no one has ever found remains of early dogs with Neanderthals. Maybe they didn't understand the idea of do-mesticating another species. How could they? It had never been done. Or maybe they were too accustomed to avoiding wolves. In any case, based on what we know now, Neanderthals did not domesticate or even tame any other species.

There may have been a few remnant populations of other archaic humans wandering around the landscape, or perhaps there were small populations of the mysterious Denisovans, a hominin named for Denisova Cave in Siberia, where their bones and DNA were found. We have so few Denisovan bones that we do not know what they looked like and we do not know their geographic range. What we do know is that Denisovans differ genetically from both Neanderthals and modern humans because genetic differences are the only way to recognize a Den-isovan. What we don't know is how different genetically two organisms must be from one another to be different species.

As the wolf was already present in this complex ecosystem, what was needed before domestication could occur was modern humans. (This may sound a little as if I think somebody or something was trying to create a dog, which is inaccurate.) The European ecosystem began to change markedly once early modern humans arrived about 50,000 years ago. One more top predator—humans—joined the guild; one more predator heightened the intense competition for resources. Every animal in the guild felt the reverberations of this change, and probably so did every prey animal in the area. The previous careful partitioning of resources and the balance of competition among predators was upset. The presence of early modern humans heightened existing rivalries for prey, water, caves, and other shelters. In the case of Neanderthals, the new rivalry applied to raw materials used in making tools. Ecologists call this type of change that resounds throughout the ecosystem a trophic (dietary) cascade. Wolves underwent a significant population drop or bottleneck as a result. This effect was probably more pronounced because the climate became notoriously unstable at this time, vacillating from dryer to wetter, colder to warmer, so wolves sought out refugia.

Early modern humans had evolved in Africa from the archaic humans who stayed behind when the ancestors of Neanderthals moved into Europe. This African group continued to evolve, develop new technology, and gather knowledge about ecosystems and other species' habits there. By about 200,000 years ago, our ancestors were fully modern in anatomy. Slowly, the distribution and geographic range of our species expanded to encompass new lands, but this was not a deliberate migration. Our ancestors wandered out of Africa—not having the least idea what a continent was or that they were leaving one for another—probably following game and avoiding population pressures in Africa.

When the first early humans reached central Europe about 50,000 years ago, they encountered a whole new set of resources and a rich predatory guild of competitors. Their survival would depend on their ability to locate those resources and to compete with the European predator guild for them. The specific knowledge of landscapes or species gained in Africa was no longer very useful, though there were certainly resemblances between the predator guild in Africa and that in Europe. For example, the cave lion of Europe did not differ so wildly from the lion of Africa, although the former had to deal with a colder climate.

The process of gathering the requisite information about the new ecosystem began most probably in the Levant—the geographic region now more commonly called the Middle East—as human distribution spread toward Europe. This is probably where our ancestors first met up with Neanderthals.

It must have been an astonishing encounter. Early modern humans were expanding out of Africa, where it is likely that all of the humans they had ever met had dark skin, dark eyes, and dark hair. We know too that the first modern humans were tall, lanky people, whose body build helped them dissipate heat. But Neanderthals had evolved separately in Europe. At least some of them were light skinned and freckled, with blue eyes and red hair. Compared with Africa's early modern humans, Neanderthals were stockier and much more muscular, which may have been in part an adaptation to colder conditions; their lifestyle also required a lot of strength, stamina, and endurance.

By the time of their encounter, both Neanderthals and early modern humans were using things such as sea shells, shaped bones, and ochre to create objects of personal adornment: think of jewelry or distinctive hairstyles or slogans on modern clothing. They may also have used body paint, tattoos, hairstyles, or scarification, as many people today do, though we have no direct evidence of this. The point of these efforts and creations was to identify yourself and your group or to mark objects you had created as yours. Among other things, the need to proclaim membership in a particular group implies that individuals met other individuals who were strangers with some frequency. From the archaeological evidence, we know that early modern humans created such symbols or markers far earlier and more often than Neanderthals. It is not clear whether either or both species had true language, but they certainly would have had at least a crude means of communicating with others of their type. So what did Neanderthals or early modern humans feel when they first met a group of humanlike beings with entirely different coloring, a different body build, unintelligible utterances, different tools, and unusual symbolic items such as they had never seen before? "Overwhelming terror" is my best guess. Deep curiosity may have followed on the heels of fear. Other anthropologists think the curiosity and excitement caused by encountering another type of human—a relatively rare event—may have eliminated the fear because true strangers would know about other areas and resources there. They had something to offer. They might even have developed tools or techniques for processing new types of resources.

Despite their cultural and physical differences, occasionally Neanderthals and early modern humans interbred. We know this from the genomes recovered from ancient bones, which seem to carry genes from two different types. The old means of defining different species—that they cannot interbreed and have viable offspring—may be contradicted here. If the two individuals can and do interbreed, but their offspring are disadvantaged or infertile, they may indeed be separate species. It is very difficult to define species boundaries in fossils. Possibly interbreeding occurred because sometimes there were so few hominins around that mates were found among strange and rather alien beings or not at all. Maybe the exotic had an appeal. Maybe I am wrong and the "other" hominins were not as terrifying as I imagine, but modern humans are often xenophobic and do not respond well to other humans who appear very different from themselves. In any case, this interbreeding left many modern humans today with a small percentage (2 to 4 percent) of genes typical of Neanderthals.

Initially, Neanderthal genes were thought to be far less common in modern Africans than in Eurasians, a point that is usually interpreted to mean that interbreeding occurred in the Levant after the first modern humans had left Africa.[6] Recent studies have indicated that the rarity of Neanderthal genes among Africans and the much higher rate of Neanderthal genes among Asians are probably artifacts of the small numbers of modern individuals that had been sampled. The first papers about a Neanderthal genome compared it with a mere handful of modern human genomes. Joseph Akey of Princeton University and a team of colleagues invented a new method for identifying Neanderthal genes that were passed into modern humans over time, in a process called introgression. With 2,504 modern genomes from around the world (including ones from five different African subpopulations, versus only a few individuals used in the original studies of Neanderthal genomes) to compare with a single Neanderthal genome from the Altai, in Siberia, Akey's team established that Africans carried 2 to 4 percent Neanderthal genes (a percentage similar to that in

Europeans and East Asians) and that nearly all of these genes were shared with Europeans. That suggests that Neanderthal genes were picked up by interbreeding when modern humans first invaded Europe, and that those genes were introgressed into African populations when those individuals migrated back into Africa. Failure to recognize this back-migration in previous analyses—as well as using inadequate samples of modern humans—seems to be the reason for the apparent lack of Neanderthal genes in Africans. Some Neantherthal genes were even advantageous and have been retained, though other genes have been lost in modern humans because of negative effects.[7]

When humans entered the new continent, they met up not only with Neanderthals and their genes but also with gray wolves, the ancestors of modern dogs. The two essential components for the domestication of wolves into dogs were in the same place at the same time.

Neanderthals soon died out, going extinct around 40,000 years ago. I have argued elsewhere that the heightened competition caused by the arrival of modern humans helped drive Neanderthals to extinction. Specifically, I have hypothesized that one way in which early humans gained the advantage in the fiercely competitive European world was by beginning to forge a long-term, mutually beneficial partnership with canids. As I see it, this relationship probably culminated in the transformation of wolves into dogs, our oldest companions. Not everyone agrees with me.[8]

I believe that by perhaps 36,000 years ago, animals that I call *wolf-dogs*—not yet fully modern like domesticated dogs, but not wolves either—became our companions. We find their skulls, their sharp teeth, their jaws, and their limb bones in archaeological sites in Europe that date back to that time, thanks to the work of my friend Mietje Germonpré, a Belgian paleontologist. Starting in 2009, she and her team found they could use a detailed statistical analysis of the shape of skulls and teeth to identify the fossils of early canids as wolves or as early dogs. I call these dog-like animals wolf-dogs, not because I think they were like today's hybrids of wolves and dogs but because they would have closely resembled wolves and only distantly resembled the dogs we know today. Germonpré often calls them Paleolithic dogs. Her technique enabled her to distinguish these animals from the ordinary wolves of the time that were fossilized at the same sites. In fact, she and her colleagues discovered that some of the ancient canid fossils that had previously been identified as wolves could be distinguished from wolves only by statistical analysis—and then they grouped more closely with specimens that until then had been considered the earliest dogs.[9]

Applying their technique to fossilized bones from several European sites, Germonpré and her colleagues have now recognized more than forty individual animals that do not fit into the wolf category and that do fit into the very early dog, or wolf-dog, category. Startlingly, modern radiocarbon dating has shown these wolf-dogs to be as old as about 36,000 years, which is much older than anyone had suspected domestic dogs to be. (Before Germonpré's work, the oldest date for well-accepted domestic dogs was perhaps 15,000 years.)

Domestic dogs that are well accepted as such were usually associated closely with human remains, and their remains were also deliberately buried, sometimes in huge dog cemeteries and sometimes with grave goods. No one would argue that a dog that has been intentionally interred in a ritual way in a specially designated place is wild or undomesticated. In fact, what I would argue is that dog cemeteries show that dogs had been domesticated long before people started interring them in a fashion similar to that of humans. After all, the very first members of the genus *Homo* did not bury each other, as far as we know. Thus, I believe that the buried dogs were treated as if they were human, or almost human, because they had been living intimately with humans for a long, long time.

The wolf-dogs were not the same as modern domesticated dogs; they may not even have been directly ancestral to today's domesticated dogs. One way to explore the question of relationships is to analyze mtDNA, which is carried only by the maternal line. Each cell carries many more copies of mtDNA than of nuclear DNA, so it is easier to retrieve mtDNA than nuclear DNA from ancient or degraded specimens.

The mtDNA extracted from the bones of the wolf-dogs identified by Germonpré's team does not match that of any modern dog, so far, but it does match the mtDNA of other wolf-dogs Germonpré has identified by statistical analysis of shape. Does this mean these wolf-dogs are not ancestral to modern dogs? Possibly, but not definitively. Mitochondrial lineages go extinct rather often because of random events such as a female's failing to have female offspring. Every time a mother has no daughters that survive long enough to pass the mtDNA on to the next generation, that individual lineage goes extinct unless another individual—perhaps a sister—carries the same mtDNA and has female offspring that do survive long enough. The hard truth is that nearly all mtDNA lineages go extinct over time. Thus, the fact that all of the maternal mtDNA lineages from the few wolf-dogs went extinct—or have not yet been found among living dogs—does not mean that the Paleolithic dogs had no surviving offspring.[10]

The inability to find wolf-dog mtDNA among those living dogs that have been tested may mean that the wolf-dogs were not the ancestors of modern dogs, and it may mean that these wolf-dogs were the results of an early and unsuccessful attempt at domestication that did not succeed in the long term. But this evidence may also mean that one particular wolf-dog lineage went extinct, which happens fairly often. What is convincing to me is that the fossils that were identified as wolf-dogs based on the shapes and proportions of the skulls all share the same mtDNA. They also shared similar diets, and those diets were different from the diets of the humans and the wolves at the same site. In checking the chemical composition of the bones of wolf-dogs and the bones of wolves, Hervé Bocherens of the University of Tübingen and colleagues found that the wolf-dogs fed mostly on reindeer meat, while wolves and humans from the same site ate substantial proportions of mammoth.[11]

We can say that the wolf-dogs were both different from the wolves at the time and seem to form a coherent group in appearance, genetics, diet, and behavior. Starting at 36,000 years ago, wolf-dogs have been found only in European sites made by early humans who belonged to the culture often called Gravettian. Nearly all of these sites with wolf-dogs contain the remains of extraordinary numbers of dead mammoths, which are rare in archaeological sites from earlier times. That finding suggests that the wolf-dogs may have been fed reindeer meat by humans, perhaps while tethered or penned up so that they didn't steal food humans needed or preferred. If true, such provisioning is a good indicator of an intimate and long-lasting relationship between humans and wolf-dogs.

A recently published article by Germonpré and a new group of collaborators tested these conclusions using a well-established technique not before applied to questions about the domestication of dogs. Dental microwear texture analysis (DMTA) is based on the observation that objects chewed or consumed by an individual in the last few days or weeks of life will damage

the teeth of the individual according to the material properties of the chewed substance. For example, eating nuts and
fruits will produce a different pattern of microscopic wear from
eating raw meat or vegetable mush. The question was: Would
this approach separate out the same two groups of canids
(wolves and early dogs) that the shape analyses of the skulls
and jaws did? And would the chemical analysis of the bones of
individual specimens separate the canids into the same two
groups, thus proving that the groups had an ecological validity?
In particular, the DMTA would help confirm or refute the idea
that the early dogs were being provisioned by people (or were
scavenging remains of human meals). If that were the case, the
DMTA ought to show more damage from chewing on more hard,
brittle bones than from eating meat because humans were unlikely to give meaty food remains to wolf-dogs. The parts of
the carcasses that humans had difficulty eating were likely to be
bony, with only scraps of meat or marrow remaining.[12]

The team selected a particular region of the jaw, where the
molar teeth are known to be used for crushing bone and other
hard materials, to examine closely. The microwear features
were measured automatically, and the overall roughness of the
texture of the selected area of each tooth was compared statistically with the textures of the other teeth. The microwear
textures of the teeth showed a distinct difference between the
two groups of canids, with more evidence of consumption of
hard, brittle substances among the early dogs. The confirmation of the hypothesis by yet another technique provides strong
evidence that the fossil site at Předmostí preserves two eco-
logically distinct groups of ancient canids. The consilience of
several radically different techniques of analysis is convincing.

If early humans were living with, working with, and possibly
feeding these wolf-dogs about 36,000 years ago, as several

different lines of evidence indicate, our ancestors had made a discovery with staggering implications. They had learned that cooperating with another species and living intimately with it meant that humans could "borrow" the other species' abilities—in the case of dogs, abilities such as a keen sense of smell, the stamina to run swiftly and nearly tirelessly after potential prey, and good eyesight—without having to evolve these traits.

Angela Perri of Durham University captures the dog-human relationship eloquently in calling hunting dogs a specialized, extractive technology used for exploiting terrestrial game. In Japan, the Jōmon hunting-gathering-fishing culture from about 12,500 to 2,350 years before present produced more than 110 individual dog burials along the eastern side of Honshu Island. Perri reviewed both the Japanese and English publications about these sites and examined the bones themselves for signs of butchery and cause of death. She argued that these dogs were specialized hunters of wild boar and sika deer in temperate forests, where hunting such animals was difficult. She speculated that elevation of dogs to near-human status was tied to special abilities or heroic deaths during hunting expeditions.[13]

In a packed ecosystem such as that of Ice Age Europe, full of many different predators, working with another species to get more food with less danger and less effort would have conferred a very significant advantage. This advantage may be directly tied to the sudden appearance of a number of Gravettian sites in central Europe with the remains of many mammoths; earlier sites had only a very few mammoths. Something changed to improve our ancestors' ability to hunt very large animals. There are no big changes in the tools or weapons of this time. I hypothesize that working with wolf-dogs might have been the key. This scenario relies on known behaviors of wolves. If wolf-dogs were similar to wolves, then they could find, harass, and hold a mammoth or other large prey animal in place more effectively than humans could. Wolves in Yellowstone National Park do this now with bison, the biggest prey species in the park. Wolves are often injured in such situations because pulling down and killing an exhausted and injured animal is dangerous. But if the wolf-dogs surrounded a mammoth and cooperated with humans, then the humans could approach and kill the cornered animal using distance weapons, such as thrown spears or arrows. Wolf-dogs and humans could then safely camp near the kill site and guard the carcass from the other predators that would try to steal it, or to steal from it. Moving a whole dead mammoth would pose a formidable problem.

If humans benefited from the intimate association with wolf-dogs by being able to take large prey, wolf-dogs benefited from the intimate association with humans by the lesser danger encountered in killing large animals and by obtaining meat more reliably. Living and hunting cooperatively with humans also offered wolf-dogs protection from competing predators, including resident wolf packs into whose territory they had strayed. Both human and wolf-dog probably also benefited from the emotional bond that is a major attraction of domestic dogs today.

Leaving aside the nuances and details for the moment, this summary affords a brief glimpse of what may have been occurring in human and dog evolution in Europe between 50,000 and 40,000 years ago. Humans perturbed the ecosystem and upset the behavioral and ecological balances that had evolved over millennia. One dramatic result was the extinction of Neanderthals, who were unable to survive the heightened competition in a period of climatic deterioration; they died out by about 40,000 years ago. Within perhaps another 10,000 years, many of the other Ice Age predators of Europe also met with local or global extinction: cave lions, cave hyenas, some species of cave bears, large leopards, and European dholes. Early humans and the descendants of gray wolves were the only large predators to survive this wave of extinction in Europe. That is the power of cooperation.

We might well ask what was happening elsewhere. The scenario and interpretation of evidence summarized above—and discussed at more length in my book *The Invaders*—are plausible and supported by hard evidence.[14] Was a similar process occurring in Asia and, if so, is this idea supported there by a preponderance of evidence? Could dogs have been domesticated twice? Yes, particularly because canids can usually interbreed with other canid species. What traces are left to help us piece together the complex history of human migration and collaboration? How can we decide?

TWO

Why a Dog?
And Why a Human?

WHY DOGS? Why were they the first species to form such a close relationship with humans? Why not some other creature?

To answer these questions we need to know, first of all, what a dog is. A simple way of approaching the question is to point at the one that lives with you: *that* is a dog. Frankly, though, modern dogs are so variable in their appearance, shape, size, behavior, and temperament that pointing is not a useful answer. Dogs vary in size from one-kilogram Chihuahuas or other teacup dogs to immense breeds such as lurchers, Great Danes, mastiffs, and Irish wolfhounds, weighing a hundred kilograms or more. Alternatively, and more scientifically, you can say that a dog is a domesticated wolf, but that is almost a tautology. We know dogs were derived from wolves and most probably from Middle Eastern or European gray wolves, though some scholars think the case that dogs were derived from Chinese wolves is stronger. The genetics, the archaeology, and the paleontology provide evidence but have not yet resolved the question. Though back-breeding between wolves and dogs can and does occur, and produces puppies, wolves and dogs are not at all the same species and the puppies may not be as fertile. *Canis familaris,* the dog, is just that; a familiar animal, one of "us," a part of human society. Dogs are found wherever there are people, and they form an integral part of human society.

Wolves are *Canis lupus,* not familiar and friendly, but wild. They are intensely social, curious, and excellent hunters, but they have no particular regard for humans. The original range of various gray wolves covered much of the Northern Hemisphere worldwide, and this enormous distribution means that there are relatively minor differences between, say, wolves in Ukraine and wolves in North Carolina. The differences in appearance, size, coloring, and other genetic traits that do appear would be expected of any species with populations covering such an enormous geographic range.

Mark Derr, the author of *How the Dog Became the Dog,* maintains that "the dog is inherent in the wolf." He finds that the intrinsic nature and abilities of the domestic dog were present in the ancestral wolf. Dogs are wolves minus some qualities or behaviors, or with some of those traits exaggerated or more strongly expressed, but dogs are not wolves with extra bits added on by evolution. Most vitally, wolves lack or suppress the enormous and instinctual desire to relate to humans and to please them that is the signature of a dog. Dogs rely on humans, they relate to humans, they live with humans. Dogs love humans, in most cases. Humans define the dog's most common ecological niche. At best, wolves tolerate specific humans when they have been raised in captivity from an early

WELSH CORGI

DACHSHUND

BEAGLE

MINI POODLE

PIT BULL

3
2
1
0 FEET

HEIGHT to SHOULDER

GERMAN SHEPHERD

HUSKY

AFGHAN

SALUKI

CHIHUAHUA

GREAT PYRENEES

Domesticated dogs comprise more than 300 recognized breeds (subspecies). They are often categorized by their intended uses in hunting, guarding, or herding. Breeders select for size, shape, fur texture, coloring, and temperament; dogs have enormous variability in these features. Most breeds were formed in the past 200 years, but there are also many unrecognized types or landraces typical of particular regions.

age. In various kinds of intelligence tests, dogs tend to look to humans for assistance as often as human babies do; wolves do not. As puppies, dogs have a longer period of openness or socialization during which exposure to humans and other novel things is key; wolf puppies have this socialization period earlier in their development and for a much shorter time. For humans to successfully raise wolf pups to live with humans, the pups must be captured very early to begin socialization. Even after they are fully grown, hand-raised wolves respond more fearfully than dogs to novelties to which they are exposed.[1]

When I try to define the personality differences between dogs and wolves, I always remember a story I was told years ago by dog trainer Chris Mason about her friend Vicki Hearne, who wrote important books about animal-human relationships. Hearne worked as an animal trainer with wolves, wolf–dog hybrids, and dogs. On one occasion, when she was traveling with a wolf–dog hybrid and her own pit bull in her vehicle, she stopped at a rather insalubrious rest stop in the middle of nowhere. The building looked shabby and small and some rough-looking men seemed to be hanging out in the parking lot, drinking and smoking. Certain that none of them would bother her or her truck, she got out and went inside to get a soda. When she came out, the men had in fact approached her vehicle. Her pit bull responded fiercely, barking, charging the doors and windows, slavering and baring its teeth, and generally threatening to tear the strangers limb from limb. The wolf–dog hybrid was calmly looking around as if bored and saying "You want her? You want the truck? I don't care! Go right ahead!"[2]

Despite working with this trainer for weeks, the wolf–dog hybrid had no instinct to protect Hearne or her vehicle, unlike the pit bull, who clearly felt responsible for the trainer's safety and possessions. A wolf is a hunter, a predator. When the human, Hearne, was hunting and training the hybrid, she was important to the hybrid, but unless they were hunting together, the hybrid had no interests in common with the human and no particular attachment to Hearne. When humans exhibited skills and attributes useful to a hunting canid, then a limited bond or collaboration was formed. Mason and Hearne suggest that what bonds a canid and another species together is a deeply felt and intensely important set of common interests and values on both sides. The pit bull in this story was deeply bonded to Hearne; it "owned" her in daily life, not only in hunting. Keeping her safe was life and death to this dog, its primary objective in life. The pit bull identified at a very deep level with Hearne.[3]

In dogs, Clive Wynne calls this deep bonding "love." Simple familiarity or being in the same place, cohabiting, was not enough for the wolf–dog hybrid; a shared morality—a belief in the other—and a common set of values would be necessary for the bond to be formed. Mason regards the bond as being based on an internal identification or goal shared between canid and person, with all the passion and fervor such feelings evoke. These descriptions are, I think, pointing to the same thing: domestication or bonding between species is behavioral, not solely physiological or genetic (though genetic changes may be necessary to fix domestication into a permanent, heritable trait). The key is that the behavioral relations between two species change, and change first.[4]

That shared morality, or belief in the other, is why the first-ever domestication happened. Who would select such a ferocious and formidable predator as a wolf for an ally and companion? Why would a wolf respect humans just because they were there or provided food? The same might be said of a dead deer. More was needed to create a meaningful link. There had to be a heartfelt commonality for such bonding.

To be clear, I don't think anyone ever consciously chose to try to domesticate wolves in the ancient past. I think early humans brought home wolf puppies because they were amusing, cute, cuddly—and because humans have an incurable fascination with animals and a deep need for physical intimacy and companionship. As the wolf puppies grew, I'm sure many were killed or driven away because they became aggressive, or defecated in the wrong places, or were simply a noisy nuisance. (These are also common reasons why badly trained domestic dogs are surrendered to shelters.) A few were better behaved and found physical intimacy and companionship a sufficient incentive to stay with humans longer, even for a lifetime. And over time, with no intentionality or plan, something emerged that I call a wolf-dog, distinguishing this sort of animal from any dog or canid we know now, including today's wolf–dog hybrids. I am not referring to proto-poodles or sort-of-collies or potential pointers. Wolf-dogs were the first step on the road to domestication that is visible in the archaeological and paleontological record. They changed our world and we changed theirs.

Wolf-dogs were not wolves, but they were much less like modern dogs than they were like wolves. Wolf-dogs participated in the beginning of a special and powerful relationship with some groups of humans. Because humans were fundamentally social predators at this time, the only way humans might form a functional and long-lasting bond with another species was if it, too, was a social predator. Perhaps as the wolves observed the humans, and the humans observed the wolves, they recognized that they each had goals and accomplishments that complemented the others' skills, and came to respect those skills. A behavioral choice that seemed good (wise, necessary, advantageous) to the wolf was judged similarly by the human, and vice versa. This made it possible to trust the other species and to work together.

Why a dog? Wolves brought stamina and speed and heightened senses and a ferocity that any human hunter might envy.

But why a human? Why did wolves choose us? It is one thing to explain why humans might have sought out wolves, but why would wolves deign to ally with humans? From a wolf's point of view, humans are probably pretty poor hunters. But humans brought distance weapons, such as spears and atlatls (spear-throwers) or bows and arrows, that both improved the kill rate and lessened the potential danger of bringing down an injured animal. What we see in the archaeological record is the sudden appearance of sites where numbers of mammoths and other huge beasts were killed without any big change in the inanimate weaponry wielded by humans. But there were canids, in fact even more canids than ever before. They were there because that is where the important action was taking place.

Ray and Lorna Coppinger, experts on working dogs, have hypothesized that the beginning of dog domestication occurred when one wolf, perhaps a lone, pregnant female feeding from human garbage dumps, self-tamed and began approaching human settlements seeking food and companionship. This scenario is certainly possible. But it is important to remember that the earliest semidomesticated dogs we know of appeared well before there were settled villages or regular garbage dumps. In such a scenario, humans are simply a food resource, much like a discarded bag from a fast-food joint. Wild canids usually avoid people and are rather elusive. Too, a lone wolf, following nomadic humans, would be in danger of fatal attack by local wolf packs when entering their territories.[5]

I am not exaggerating the risk. In Yellowstone National Park, on one unforgettable day, I watched a wolf come over a hill, probably planning to steal some of the remains of a bison that had died earlier in the morning in the territory of the Lamar Canyon Pack. The carcass was fully visible from where I was standing, though I was too far away to hear any vocalizations. The adults of the pack, with bellies round and full, were dozing in the sunshine as youngsters played in the grass. The lone wolf was from the neighboring Mollie's Pack. As soon as the leader of the Lamar Canyon Pack, a formidable female known as 06, became aware of the interloper from Mollie's Pack, the chase was on. The wolves went from sleep to full-out speed in moments. After a breathtaking pursuit, one of the wolves from the Lamar Canyon Pack caught the intruder by the tail. Though some of the gory details were shielded from my view by bushes, each of the pursuing wolves literally leaped into the fray. Clumps of fur flew. The Lamar Canyon Pack wolves were merciless. It was a battle of roughly twelve to one. The one wolf from Mollie's Pack never walked out of those bushes. *This* is what awaited a wolf who tried to scavenge from a carcass in another pack's territory.

However the first wolf-dogs emerged, it was a highly unusual circumstance. Could a human or a pack of humans offer enough to make working with them attractive? We have poor eyesight, mediocre hearing, a pathetic sense of smell, and relatively poor ability to chase prey down, and undoubtedly wolves knew that. But early humans understood the principles of hunting and worked hard to succeed. Our camps smelled of food. Our fires were warm and bright. We knew how to work together and share the spoils. And we had some almost magical means of killing an animal from a safer distance. Maybe that was enough.

THREE

What Is Dogginess?

WHAT THE FIRST DOGS had, like the last wolves, was what I call *dogginess*. Dogginess is the essence of dogdom writ large. Clearly the first dogs had dogginess, or we would not recognize them as related to modern dogs, but they certainly were not dogs as we know them now. The very first animal to exhibit dogginess cannot possibly have been a dog like those we see today, for that species had not yet evolved. Yet those first animals with dogginess were not simply wolves, either, though dogginess started from and included many of the traits of wolves. We have asked "Why dogs, anyway"? Why were dogs the first and most common domesticate? Why did dogs become our fellow travelers, our companions, guardians, playmates, herders, protectors, blankets, shepherds, and hunting aids? Why did humans invent—or stumble upon—the process of domestication to deal with wolves or dogs and not, say, deer or wild cattle or goats? The answer depends upon understanding both what the nature of dogginess is and what it meant to domesticate another predator and to live intimately with it. I will try to show what this extraordinary process looked like in real life, insofar as we can deduce that from the evidence. Understanding dog domestication explains why we form such deep emotional bonds with dogs. In fact, I argue that the emotional bonds come first. The journey with our doggy traveling companions has altered both us and them—physically, behaviorally, genetically, and emotionally. Together, humans and dogs have embarked on and succeeded at all kinds of crucial endeavors.

What exactly is dogginess? As a trait that separates larger canids from other mammals, dogginess has something directly to do with a general shape and build, a particular style of life to which canids are admirably adapted. In wolves—which despite their overall appearance are canids but not dogs—there is nonetheless quite a bit of dogginess. There is the need to know, to observe, to be aware, to communicate with others of the group or pack, to chase, to track, and to kill prey. In dogs per se, dogginess includes a desire or a drive to be part of and communicate with a group that includes humans. Communicating may mean living with humans or it may mean cooperating with humans; neither extreme is necessary to be a dog in the greater sense of the word. In modern dogs, this trait is often expressed as a desire to be with humans, physically and emotionally, and to participate in humans' lives. In some cases, this intense desire is manifested as separation anxiety on the dog's part—a very common ailment in which the dog is deeply frightened and panicked when separated from its human or humans. Separation anxiety shows up as ceaseless barking or howling when alone, destruction of whatever the dog can get hold of, defecation or urination in inappropriate places, and sometimes self-mutilation from excessive licking or chewing or attempts to escape by damaging doors, windows, or other potential exits. Extreme separation anxiety is a very common trait among dingoes living as pets and one that makes them difficult to maintain.

Dogginess is also a certain intrinsic alertness to sound and smell, to intruders or strangers, to cause and effect, to slight movements or intangible changes in atmosphere that assist animals that are fundamentally hunters to survive—even those dogs that are now amusing companions and not hunters. It is also a curiosity about the world, a cleverness, an interest in solving puzzles or exploring the little known. Dogginess is also, often, a love of movement, of action, and of extreme smells. A lot of dogs love to run and need to run. Not all canids show all aspects of dogginess, but domestic dogs, wolves, dingoes, and, I would argue, even the earliest identified dogs do and did show some of these characteristics. The widespread nature of dogginess leads me to suggest that there is more than one way to be a dog. In other words, dogs are not only domesticated wolves, but all doglike canids I know of—foxes, coyotes, and others—exhibit significant dogginess.

Dogs also bond strongly to their pack, which forms their family. In modern dogs, this tremendous ability to form bonds—with other dogs, or humans, or sheep, or penguins, or other animals to which they are exposed when young—is just about a defining trait. "Loyalty," which is often cited as characteristic of dogs, refers in large part to the strength with which a dog can bond with another. Clive Wynne has explored this ability to bond in his book *Dog Is Love*. He argues that this ability to bond is an intrinsic quality of dogs and one of their defining traits. He is probably right. Synthesizing his own research on dog psychology with genetic findings by Bridgett vonHoldt, Wynne has come up with a fascinating and powerful idea.[1]

VonHoldt led a study of the genomes and single nucleotide polymorphisms (SNPs, pronounced "snips") in canids, using 912 dogs and 225 gray wolves as her sample. SNPs are locations in a genome where one of the four nucleotides (adenine, cytosine, guanine, and thymine) has been substituted for another and the substitution is a common variant in the species. The pattern of different SNPs is what makes each individual's DNA unique. VonHoldt hoped that looking at patterns of SNPs in dogs might reveal where that genetic variability arose. Only a few of the eighty-five breeds she genotyped showed evidence of much admixture with gray wolves, though it is known that modern wolves do crossbreed with dogs. Mostly the genes seem to cross from dogs into wolves and not vice versa. Clustering each breed with its nearest (genetic) neighbor revealed a group of dogs that are highly divergent from others: these are the Basenji, Akita, chow chow, Afghan hound, saluki, Samoyed, Siberian husky, Alaskan malamute, dingo, Canaan dog, Chinese shar-pei, New Guinea singing dog, and American Eskimo dog. Each of these breeds has been considered "ancient" because historical information suggests the breed formed more than 500 years ago, meaning it is probably a primitive breed. As one of the unanticipated findings of this massive study, the authors remark, "Furthermore, we observed a single SNP . . . located near the *WBSCR17* gene responsible for Williams–Beuren syndrome in humans . . . , which is characterized by social traits such as exceptional gregariousness."[2]

Williams–Beuren syndrome (WBS) in humans is not a widely known condition, but it is one of those for which the genetic component is well understood. At least twenty-seven particular genes have been implicated in WBS, which results in a characteristic facial type often called "elfin," some cognitive impairment, and possible heart or kidney problems, depending upon how many of the relevant genes are affected. WBS patients are notable as much for their lack of shyness and outgoing personalities as anything else. They are drawn to make social connections with other people.

Domestic dogs show alterations in the genetic code of up to six of the WBS genes and, like humans with WBS, are very likely to initiate social contact with strangers. This is the trait that Wynne calls "love" in dogs: intense bonding to humans, or sometimes to another species. However, the precise mechanisms in dogs and people differ. In humans, there is a deletion of approximately twenty-seven genes, and the regions of the genome bounding the deleted area show diminished expression. In other words, the deletions block the production of some of the substances the body usually produces. In dogs, the mechanism seems to be different, but some of the effects are similar. Dogs have mobile element insertions that block or reduce the effects of four of the genes in the canine WBS region; these insertions thus may be responsible for the greater friendliness toward humans, and lessened aggressiveness, in domesticated dogs compared with wolves.

Brian Hare of Duke University has hypothesized that selection for more tolerance toward and better communication with humans formed the basis of dog domestication, an idea similar to that propagated by Ray Coppinger.[3] Bridgett von-Holdt and her team's genetic studies may have discovered exactly how that tolerance and communication occur. How exactly one fosters this sort of bonding between two species remains mysterious.

To be fair, not all animals we consider domesticated have such a deep ability to bond with humans—and tomatoes, corn, and wheat don't seem to have that ability at all. A few animals can develop such bonds given the right partners or opportunities, but plants basically don't have emotional ties to other species. In my view, plants may be cultivated, they may be selectively bred, but they can't be domesticated.

Dogs and humans evolved together and domesticated each other. What does this mean? It means that, over time, each species has been genetically modified in ways that enhance their communication and thus facilitate their living together. This might be called coevolution. I see it as developing a common behavioral language.

FOUR

One Place or Two?

THE POSSIBILITY of human and dog coevolution in Europe appears to be fairly straightforward and well supported by evidence. Domestication requires at the very least the presence of canids and humans in the same place at the same time, with the same interests. But that is not the entire story, which is complex. Consider the situation in Asia at that time, which also had both wolves and humans.

One means of trying to unravel the complexity of the first domestication relies on genetic studies. Sometimes these studies have been highly useful; sometimes they have been highly confusing.

The most perplexing and poorly documented element in the story of the first dog was completely unknown until 2008, when genetic analysis of fossil bones revealed something quite unexpected. About 50,000 years ago, it was not only modern humans and Neanderthals that were present in the Old World—there was a third, long-unsuspected hominin (humanlike creature) as well. This assertion is based almost entirely on genetic analysis of a scant handful of fossils from a single site, Denisova Cave in the Altai Mountains of southern Siberia. Tools and ornaments typical of early modern humans and of Neanderthals had been known from various levels at Denisova since excavations made in the 1970s. These artifacts were taken as evidence that both early modern humans and Neanderthals inhabited the cave at various times, but there were no modern hominin bones, and documentation of the early excavations was sometimes not up to modern standards. But in 2008, ongoing excavation at Denisova led by Michael Shunkov and Anatoli Derevianko of the Russian Academy of Sciences in Novosibirsk revealed the first of a set of humanlike remains. After another ten years of work, the result was three teeth, a small bone from the finger of a juvenile individual, a toe bone, and a nondescript piece of a broken long bone. This is pathetically scanty. Though the fossils were clearly hominin, they were not distinctive or nearly complete enough to tell the scientists involved which hominin they represented.[1]

Shunkov and Derevianko turned to a top genetics group that included Svante Pääbo and Johannes Krause of the Max Planck Institute and David Reich of Harvard Medical School. The geneticists were able to extract mitochondrial DNA (mtDNA) from two of the teeth from Denisova and both mtDNA and nuclear DNA from the finger bone. The ancient mtDNA samples were compared with the mtDNA from five modern humans (one Chinese person, one West African, one South African, one French person, one Papuan), six Neanderthals, a bonobo (pygmy chimpanzee), and a chimpanzee, which is a very modest but wide-ranging sample. However, the results were startling. The mtDNA of the new fossils matched neither that of modern humans, nor that of the Neanderthals, nor that of the bonobo or chimpanzee. The nuclear DNA from the finger bone did not match the Neanderthal genome from Altai, Siberia, nor was it like the nuclear genomes of the five present-day humans. It is a precarious leap to assume that the genomes of five modern humans will adequately sample the genetics of the more than 7.8 billion humans alive today, but the team made that leap. In contrast, genetic analysis of the toe bone and long bone fragment from Denisova yielded Neanderthal mtDNA. The team's findings showed that apparently some of the bones from Denisova were from neither Neanderthals nor modern humans, but from another type of hominin. Who or what was it?[2]

This finding posed a curious problem. Nobody expected to find another new and wholly unknown species of hominin in Siberia. In fact, nobody really knows how genetically different two individuals have to be in order to belong to different species, nor do they know how many genomes have to be sampled to give an adequate idea of genetic variability within a species. That problem arises in part because the answer depends on which specific genes are involved. Nobody could give a formal description of the unique anatomical features of this new hominin—a task that is part of the standard for naming a new species—because neither the two teeth nor the finger bone revealed any unique anatomical features. Nonetheless, the geneticists were confident of their findings: they had something new. With great fanfare, they announced the discovery of a previously unknown species of hominin. Because there is so little fossil material, and what is known is so uninformative, these unknown hominins have been nicknamed Denisovans rather than being given a new, formal, taxonomic name.

Remember, though, that the assertion that Denisovans are new is based on comparison with only a handful of precious

Neanderthal genomes and with a scant fifty-four human mtDNA genomes and five human DNA genomes. Are the comparative samples big enough to make such a judgment with total confidence? Maybe, maybe not. A larger number of Neanderthal or human samples could indicate that Denisovans are simply unusual Neanderthals or rare modern humans.

Niccolo Caldararo of San Francisco State University has been outspoken about these studies, arguing that there are large amounts of contamination and degradation that have been overlooked in the genomes of Neanderthals and Denisovans. Some contamination with modern mtDNA has also been noted by others in the original mtDNA Neanderthal sequence. Caldararo suggests that incorrect alignment of Denisovan, human, and Neanderthal sequences for comparison has probably produced erroneous degrees of difference between the genomes. His perspective is not widely accepted in the field, yet he makes some strong points about the difficulty, and thus the possible unreliability, of extracting accurate ancient genomes from fossils.[3]

If we postulate, cautiously, that the genetic differences between Neanderthals, modern humans, and Denisovans are both more or less correctly identified and great enough to suggest a species difference, what do they mean? Fascinatingly, both Neanderthals and Denisovans left a few pieces of their skeletons in the cave, but modern humans did not. When were *they* there? Dating techniques applied previously to the Denisova Cave remains either did not incorporate modern decontamination techniques or are of questionable reliability in caves. However, geochronologist Tom Higham and his team from the Oxford Radiocarbon Accelerator Unit were able to re-date a number of burned and cutmarked animal bones from the cave—bones clearly modified by hominins—as well as to date a new fragment of Neanderthal limb bone from the cave. The new limb bone specimen was exceptionally well preserved and allowed the Oxford laboratory to say that it is at least 50,000 years old, and maybe more. (Radiocarbon dates are not accurate for remains older than 50,000 years ago, even when stringent decontamination procedures are used.) There is no piece of modern human bone from Denisova Cave to try to date.[4]

A partial jaw from a high-altitude cave, Xiahe, in Gansu province in China, also appears to be from a Denisovan. The fossil had been found in 1980 by a Buddhist monk who had gone into the cave to pray, but it only came to the attention of paleoanthropologists in 2010. Unfortunately, the fossil's precise position in the cave when it was found is unclear. Its antiquity is estimated at 160,000 years ago, based on analysis of a carbonite crust adhering to the fossil, assuming the crust was formed at the time the fossil became buried. Geneticists were unable to extract mtDNA from the fossil, but collagen proteins extracted from the jaw (collagen is a key protein in bones and teeth) resemble those found in Denisovans. Thus, both the antiquity of the jaw and its identity are questionable. However, the second molar tooth in the Xiahe jaw appears to have three roots, not two, which is a trait found in Asian-derived populations, including Native Americans. The adjacent first molar tooth is missing, but was lost after death. Its socket suggests it was three-rooted also.[5]

Questions have been raised in a recent paper by G. Richard Scott and colleagues, who point out that the three-rooted molar trait occurs in high percentages in Asian lower first molars, not second molars. So the anatomy linking the Xiahe jaw to Denisovans may not be as secure as it might have seemed.[6]

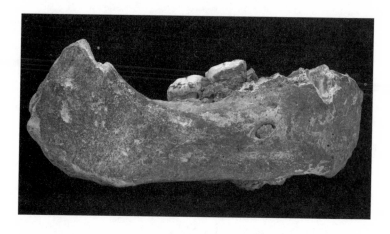

This ancient jawbone was found in Baishiya Karst Cave on the Tibetan plateau in Xiahe County, China, in 1980, but came to the attention of paleoanthropologists only in 2010. Analysis of protein in the fossil suggests it is the same species as some of the fossils from Denisova Cave in Siberia, making this jawbone the only Denisovan fossil found outside the Siberian cave.

Another unanticipated finding was that some modern humans still have some Denisovan genes, notably *ETAS-1*, which is also found among living Tibetans, who use this gene to help them adjust to low oxygen levels related to living at high altitude. However, Denisovan genes are also found in New Guineans, Bougainville Islanders, Aboriginal Australians, and a few groups from Island Southeast Asia, as well as among some Indigenous peoples of Amazonia in South America, Tibet, and parts of eastern Eurasia. This is an extremely peculiar geographic distribution, to say the least. The distribution straddles enormous geographic barriers, including the Pacific Ocean and several mountain ranges. Further, few of the modern groups identified as carrying Denisovan genes live at high altitude, so the survival of those genes makes no particular sense (nor is Denisova Cave at high altitude). Can science support the assertion that Denisovans are not the same species as modern humans, when modern humans carry their genes?[7]

Frankly, a map of who lived where some 50,000 years ago is puzzling. We know that early humans expanded out of Africa and that some encountered Neanderthals, probably interbreeding in the Levant. Some humans retained a small percentage of Neanderthal genes as they continued to expand north to Europe, then east to northern Asia, where they may have met Denisovans and interbred. Some humans almost certainly made a reverse migration back into Africa, carrying genes with them. Contrary to what was previously thought, we now know that modern Africans do carry a small percentage of Neanderthal genes. Other humans interbred again, retained a few more genes, and then probably expanded into Beringia, the now-submerged area that connected Asia to the Americas when sea levels were lower than they are now. During this expansion into the Americas, a few Denisovan genes were carried along. Most paleoanthropologists think this journey occurred about 30,000–20,000 years ago at the earliest, but this point is contentious. Denisovan genes were apparently lost from most of the lineages ancestral to today's modern humans, but not all. Genetic analyses made in 2015 by Pengfei Qin and Mark Stoneking found that the extent of Denisovan genes in different modern groups correlated more closely with the percentage of New Guinean ancestry than with the percentage of Indigenous Australian ancestry, though Indigenous New Guineans and Indigenous Australians are closely related. Before about 8,000 years ago, New Guinea and mainland Australia were connected by a land bridge rather than being separated, as they are today, by the Torres Strait. Did most Denisovan genes arrive in the supercontinent known as Greater Australia

(New Guinea, mainland Australia, Tasmania, and other islands that were close by) after Australia became an island continent? Or were they carried into Greater Australia by people from eastern Asia?[8]

What would that mean? Why weren't more Denisovan genes carried along by all this intermixing and migration? The finding certainly suggests that most Denisovan genes were not advantageous to modern humans, based on their current location, and possibly that the human-Denisovan or Neanderthal-Denisovan hybrids were less fertile. This would cause the carriers of the Denisovan genes to die out over time.

Judging from the mtDNA, Denisovans were closer genetically to Neanderthals than modern humans are to Neanderthals, making Denisovans a sister group of Neanderthals but not their ancestor or descendant. The number of different sorts of humans in the past is rather confusing.

But where did Denisovans come from? This is a point where the story of human migrations gets tantalizingly unclear. Awkwardly, Denisovans are virtually invisible until they got to Siberia at about 50,000 years ago, except for the poorly dated partial jaw in a remote cave on the Tibetan plateau that might be 160,000 years old. Denisovans must have been somewhere prior to 50,000 years ago, but we do not know where. If they were on the Tibetan plateau of China, that would confirm the tentative identification of the Xiahe jaw as Denisovan. Because the handful of remains that have been studied genetically are not anatomically distinctive, we don't know how to recognize Denisovans older than 50,000 years until they have been genotyped. Such an analysis of the Xiahe jaw is impossible with current techniques. There is fossil evidence of hominins in Asia before about 60,000 years ago, at places such as Maba, Xujiayao, Dingcun, and Jinniushan in China, but we don't know if these fossils represent Denisovans or some other archaic species.

Somehow people carrying Denisovan genes went southeast from the cave in Siberia—again remaining invisible in the known fossil and archaeological record—and after another 45,000 years, they moved onto the highland plateau of Tibet. Domesticated yaks are seen as crucial to year-round human settlements in that habitat, yet yaks were domesticated only about 5,000 years ago. Yaks are so important because yaks provide more yaks, as well as meat, milk, skins, haulage or power, fur or hide, and dung for fuel in a cold, harsh, and sparsely treed environment. Before yaks were domesticated, humans may have spent the coldest and most difficult months in the more protected valleys and lowlands, living on the plateau only seasonally. In addition, some Denisovan genes also help modern Tibetans live easily at high altitudes, where the lower air pressure means that it is more difficult to keep blood sufficiently oxygenated. So far, the same genes that help modern Tibetans deal with lower levels of oxygen in their blood have been found nowhere else among the samples tested except in two lowland Han Chinese individuals.[9]

The Han Chinese genes were sampled by the 1000 Genomes Project, a collaborative, international effort started in 2008 to create a comprehensive catalog of human genetic variability by sampling at least a thousand individuals from around the world. Out of 7.8 billion humans, a sample of a thousand individuals seems pretty paltry. Inevitably, the resultant database omits many ethnic groups, often peoples from Australia, Oceania, and New Guinea. Many groups that are sampled are represented by very few individuals, such as the two Han Chinese. Sampling issues are doubtless why these Denisovan genes show up in so few individuals other than Tibetans in the 1000

Genomes Project database. Because China had a population of 1.386 billion as of 2017, with 553 recognized ethnic groups, it is clear that much better sampling is needed before we can confidently say that such genes are rare outside inhabitants of the Tibetan plateau.[10]

Why did Denisovans have high-altitude genes, anyway? Denisova Cave is about the only place we know for sure that Denisovans lived, and it is not at high altitude. Why were those genes preserved? Where were Denisovans between 50,000 years ago, when they lived in Denisova Cave, and about 5,000 years ago, when they apparently moved into highland Tibet with yaks? If Denisovans lived on the Tibetan plateau 160,000 years ago, as the Xiahe mandible suggests, how did they survive without yaks or their dung to use as fuel for fires?

Those high-altitude genes would not have been useful or advantageous to the ancestors of modern people with some Denisovan genes who went on to inhabit New Guinea, Australia, and Bougainville Island, or to particular groups in Oceania, such as the Mamanwa of the Philippines. As Denisovans (or their genes) moved far south and east from Siberia, they apparently left few surviving relatives in central Asia.

Another important question whose answer is as yet unknown is where the modern humans who ended up in Greater Australia picked up Denisovan genes. A coastal route along the southern coast of India to what is now Indonesia, and then on to Greater Australia, is often postulated, but has little fossil support. (Neither does any other route.) We do know that people could get to Greater Australia only by boat, which makes it more likely that the First Australians came from a coastal population than from an inland one. All proposed routes involve crossing Wallacea, the faunal and floral zone that lies between Asia (Sunda) and Greater Australia (Sahul). Unfortunately, archaeological sites are rare in Wallacea.[11]

Recent calculations of the least cost (to humans) of various routes toward Australia after crossing Wallacea show substantial support for a pathway called the northern route, traversing a series of islands, including Sulawesi, that emerged or were enlarged when sea levels were low. The most probable landing point in Greater Australia is Misool Island, which lies to the west of New Guinea and was at the time the westernmost part of Sahul. As the authors Shimona Kealy, Julien Louys, and Sue O'Connor remark:

> The combination of shorter crossing distances and continuous absolute intervisibility makes this route the most likely option in all modeled scenarios. This does not rule out the possibility that other routes were used by the early inhabitants of Wallacea to travel between islands and onto Sahul. Our model nevertheless suggests a northern route from Sulawesi, and through either Obi or Seram, with a landing point near the New Guinea Bird's Head (i.e., Misool) was the easiest based on the variables examined, and thus likely the earliest route taken through Wallacea and into Sahul.[12]

Much later, other people traveled from northern Asia across the Bering Strait into the Americas, bringing Denisovan genes with them. The genes could then have been transmitted southward into South America without leaving discernible genetic traces along the way (more invisible evidence). This does not necessarily mean that Denisovan individuals themselves moved

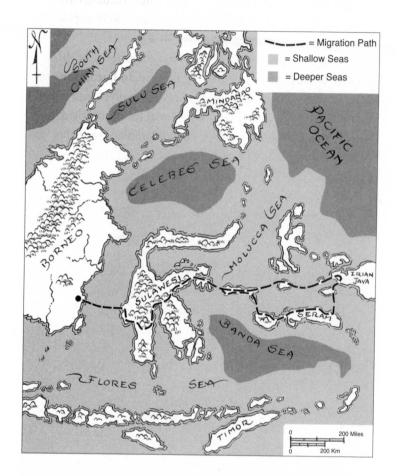

To reach Greater Australia, early modern humans must have traversed various bodies of water in Wallacea. Modeling that incorporates information on island intervisibility, rising or falling sea level, bathymetry, and paleocurrents strongly favors the northern route, shown here.

as far as the Americas, and they probably didn't. Genes can "move" through long chains of local interbreedings among a series of populations, which seems the most likely explanation to me until we have new evidence.

Confusing as the interbreeding and migration of different hominins is, we can say one thing for sure. In Europe and Asia about 50,000 years ago, there were both wolves and modern humans, as well as archaic humans. The domestication of wolves into the first dog could have taken place on either continent, or both.

But did it?

FIVE

What Is Domestication?

To ANSWER THE QUESTION of whether or not the first dog evolved in Asia or Europe, we need to go back and create a good working definition of domestication. "Domestication" has a very specific meaning. The term is derived from the Latin for "dwelling" or "house": *domus*. In its broadest sense, domestication is the process of rendering an animal or plant suitable for or amenable to living in the domus, for being a member of, and living intimately with, the family.

Even in this general sense, the precise meaning of domestication is elusive. Are plants domesticated? Certainly some of them are spoken of as domesticated, as needing deliberate care and cultivation, and sometimes fertilization, by humans and, conversely, as having been genetically modified through human selection to have traits considered desirable. I am not talking about the relatively recent process of genetically engineering changes to plants; these modified products, such as soybeans, are known colloquially as GMOs (genetically modified organisms). Selection has been carried out for millennia by hunters, gatherers, foragers, gardeners, farmers, and breeders of various species through old-fashioned means, not in the laboratory. If you want, for example, violets with white stripes, what do you do? You try to nurture the seeds of those that show white stripes and pull up the ones that don't, until you always get striped ones (if you ever do).

We can understand the general principle of selecting or choosing the most desirable plants—those that yield the most food under particular conditions, for example—but the practice of selection is somewhat paradoxical. The individual plants that produce rich fruits or seeds or tubers are the ones you would most want to eat—and those are the very ones you must save for the next planting season. Which is the most practical strategy? Why did people start saving the best seed? It is an awkward conundrum. As the late Brian Hesse wisely observed in his studies of early domestication, people who are short of food, even starving, do not save food for next season or next year. They simply try to live until next week.[1]

The habit of saving seeds for another day must have arisen in relatively good times, when food was plentiful enough to keep some for the distant future. This implies that the motivation for domestication is not to ensure a stable food supply because undertaking the initial process of domestication makes sense only if you already have enough food. Plant domestication seems to be about improving the plant species in the long run. But you really don't care if the plant is happy to see you or plays nicely with the children.

What is more, strictly speaking, domesticated plants— crops—do not exactly live with humans or in the home. In fact, because some of them, such as nuts and fruits, grow on trees, and most require sunlight, they could not possibly live indoors.

Domesticated plants certainly do not participate in family life in any active way, though their needs and locations may shape the seasonal and daily round of activities and the locations of settlements. They don't join the family. There is an odd sort of remote intimacy between crops and those who harvest or farm them.

The more you ponder the domestication of plants, the fuzzier the concept of "domesticating" them becomes. The earliest farmers or gardeners did not know enough about the mechanics of reproduction or genetic inheritance to know how to get a particular plant to fertilize some other particular plant and produce bigger corms, or juicier fruits, or nonexploding seed heads (which are easier to harvest), or tubers that were richer in carbohydrates. Domesticating plants was not a matter of learning which individual plants were friendliest or least aggressive toward people. And yet, over time, wisdom accumulated, sometimes accompanied by good luck, and humans did find out how to alter some plants' genetics to foster a more desirable outcome. This discovery is often spoken of as the Neolithic revolution or the dawn of agriculture. It is generally thought to have occurred around 11,000 years ago. Agriculture as an organized system of growing food transformed at least some people who had traditionally hunted, gathered, and foraged for their daily food—mobile people living off the land—and turned them into more sedentary farmers, tied to fields and villages and dwellings.[2]

The Neolithic revolution was not a win-win proposition at the outset. Several studies have shown that early farming peoples experienced a decline in their general health because they often had monotonous diets based on a very few staple resources. Having a narrower range of staple foods meant that those people were more vulnerable to normal variations in weather, such as too much or too little rain, or too hot or too cold or too short a growing season; and of course there were plant diseases, which spread easily when a whole field is planted with a single species. Growing crops also caused humans to live in more permanent settlements, which exacerbated problems with sanitation, water supply, and human crowd diseases. Though farming supported more people living in higher densities than hunting and foraging, it also created perfect conditions for the spread of contagious diseases and parasites and for recurrent episodes of starvation in bad years.

And then there was warfare. Among nomadic foraging and hunting peoples, disputes are often settled by one group moving away from the other. But clearing and fencing fields, planting and tending crops, and building storage facilities takes a lot of work, so people begin to defend territories—or to raid others' territories when times are bad and their own crops fail. Excess foods, such as the seeds for next year or the vegetables saved for winter, could be stolen during a raid. Abandoning a cleared or planted field and a store of food is an expensive proposition, much more risky than simply shifting your hunting to another area when game gets scarce or your brother-in-law becomes annoying.

As best we know at present, the domestication of plants began about 11,000 years ago with fig trees, emmer wheat, flax, and peas in the Near East. At about the same time, foxtail millet was domesticated in Asia. How do we know this at all? We know it because of plant remains preserved under special conditions. Seeds can be preserved and sometimes were. Many edible plants also contain starch grains and phytoliths, microscopic silica structures that are much more resistant to decay than leaves or stems. If found, these can also be used to identify plants that were used in the past; techniques such as radiocarbon dating can tell us when this occurred.

Historically, it was often assumed that plants were domesticated earlier than animals, but modern science shows that this idea is unquestionably mistaken. There is no logical reason why it should be true. The attributes and needs of domesticated crops differ a great deal from those of hunted or gathered foods; knowing how to raise wheat tells you little about how to look after pigs. Like fields, particularly rich hunting grounds could be invaded by others and were worth defending. But many hunters and gatherers or foragers were nomadic and lived in low densities out of necessity. Staying too long in one area depleted the local prey population. Whereas agriculturalists can store crops for the future, hunters cannot store meat for long in temperate or tropical climates, though extreme cold works well to keep meat frozen. Over time, crops are more vulnerable to theft than carcasses.

Domesticating animals involves other issues. Domestic animals are not normally hunted; indeed, they are not always confined and may be free ranging. Still, domestic animals can be moved to a new area much more easily than a planted field, a store of grain, or a pile of tubers, which simply will not get up and walk to a new locale. Such animals may even transport household goods as they are being moved. Moving domestic animals is a very different proposition from moving plant foods.

So why do we use the same word, *domesticates,* to describe both plant and animal species, and a single word, *domestication,* to describe the process by which an organism becomes domesticated? I think it is a grave mistake that has been based on outdated ideas and faulty assumptions. I do not believe that a single process is involved. I argue that plant and animal domestication are radically different because the nature of the wild species from which domestication might begin is also radically different. As well as having the inherent genetic variability that causes some individuals to exhibit more desirable traits, animals must also cooperate to some extent if they are to be domesticated. Animals choose domestication, if it is to succeed. Plants do not. Like animals, plants have to have enough genetic variability to be exploited by humans during domestication, but plants do not decide whether or not to grow for humans. Animals must decide whether or not to cooperate.

Influential scientific discussions about the domestication of animals began in the era of Charles Darwin and his cousin, Francis Galton, during the nineteenth century. In their world, farming was a predominant way of making a living, whether the farm was a smallholding worked by a single family or an aristocrat's estate or plantation worked by people who were often tied to the land itself or bound to the landowner.

In 1865, in a prescient publication, Galton identified the key attributes of animals amenable to domestication:

> 1, they should be hardy [so they can live in captivity]; 2, they should have an inborn liking for man; 3, they should be comfort-loving; 4, they should be found useful to the savages [who he presumed domesticated them]; 5, they should breed freely; 6, they should be gregarious.[3]

He envisioned that the process of domestication happened by slow selection for the most desirable animals, followed by the inheritance of the traits deemed best, so that the irreclaimably wild members of every flock would escape and be utterly lost; the wilder of those that remain would assuredly be selected for slaughter, whenever it was necessary that one of the flock should be killed. The tamest cattle—those that seldom ran away, that kept the flock together and led them homewards—would be preserved alive longer than any of the others. It is therefore these that chiefly become the parents of stock, and bequeath their domestic aptitudes to the future herd.[4]

I particularly like the "inborn liking for man," which I think is the same trait Clive Wynn calls "love" and Brian Hare calls "friendliness." An animal suitable for domestication is not intrinsically fearful or wary of humans.

Galton also emphasized that this process was not designed to fulfill a preconceived intention but happened many times with varying success, by accident. Over time, repeated failure to care for individuals with undesirable traits or actually culling them will produce significant genetic change in the population.

Despite knowing nothing of the mechanisms of inheritance, Galton, and Darwin after him, identified many of the important factors that influenced domestication and breeding. In his 1868 book *The Variation in Animals and Plants under Domestication*, Darwin carefully distinguished between unconscious selection, "which results from every one trying to possess and breed from the best individual animals," and methodical selection, which is carried out "with a distinct object in view, to make a new strain or sub-breed, superior to anything of the kind in the country."[5] His point is that selecting individuals with no predetermined intention of making a long-range improvement in the species is the most probable initial step in domestication. I would agree. There is no possibility that the first person to participate in domesticating an animal could have foreseen the outcome. As Galton observed, at least the first domestication was most unlikely to have been undertaken with any specific aim in mind. Although hunting with dogs is often demonstrably more successful than hunting without dogs among modern peoples, that situation cannot have been envisioned or imagined when the domestication of wolves into dogs first began. Whether dogs were domesticated *in order* to function as a hunting aid or not is another matter. Galton and Darwin agree to a large extent with most modern thinkers about domestication and yet they both also clearly took domestic stock, not dogs or cats, as the model for the process. In the passage quoted earlier on attributes favorable to domestication, Galton refers to cattle specifically.[6]

Livestock may seem like the heart, the epitome, of domestic animals, yet they were not the first domesticate; dogs were. This is why the first dog was so important. In all of time, only two predators have ever been domesticated, whereas at least a dozen herbivores—plant eaters—have been domesticated: pigs, sheep, goats, llamas, guinea pigs, rabbits, camels, cattle, yaks, horses, donkeys, chickens, and others. Dogs were the earliest domesticate by many millennia; cats came much later and may have been largely self-domesticated after humans began growing and storing crops that attracted the rodents upon which cats fed. There is no good rationale for using stock animals as the prototypical animals for domestication. Possibly this confusion occurred because the two domesticates that live most intimately with us, cats and dogs, are unconsciously considered to be categorically different from stock animals. The tendency in many countries to consider cats and dogs as part of the family—as near-humans if not actual humans—is well documented. Cats and dogs receive birthday cards and cakes, Halloween costumes, toys, special beds, treats, sweaters, boots, and so on. In some sense, many dogs and cats effectively become "people" during the domestication process, whereas other animal domesticates remain animals. There is, in fact, an impressive history to the treatment of dogs as persons, going back at least 9,000 years.[7]

Dogs were not only the first domesticate, they are unquestionably the most widespread domesticate today, living in every habitat where humans live. (It could be argued that various parasites or species such as house mice are as widely spread and perhaps even more ancient than dogs, but the pairing and intimacy of parasites and hosts is a rather different arrangement. We do not consciously look after parasites or take steps to ensure their health and survival.)

What is fascinating is that even nineteenth-century scholars writing about domestication saw clearly the distinction between taming a wild animal and domesticating one—and attributed this difference to changes brought about by the process of domestication itself. However, well into the twentieth century, domestication was still regarded as the foundation of civilization, which explicitly involved an agricultural lifestyle. I suspect that this view is based on Eurocentrism, at least in part. Civilization was manifest in practices such as living in permanent settlements, developing some form of government, developing "luxury arts" such as weaving and pottery, "specialized labor . . . [and] sophistication of religious and ritual belief and practice," and above all, the accumulation of surplus so that obtaining sufficient food every day was not such a time-consuming activity. This was how civilized Europeans did it, anyway. Until very recently, the transformation from a hunter-gatherer lifestyle to that of farmers or ranchers was believed to require the domestication of plants and the simultaneous origin of agriculture. The old story ran thus: Humans domesticated grains, planted and tended fields, reaped the produce or grain, and then domesticated stock animals to eat the otherwise useless stubble in harvested fields. Later, stock animals could be used as power for plowing or transport of goods, as a source of milk or hide, and to produce more livestock, until the end of usefulness came and the livestock could be eaten.[8]

Unfortunately, this "self-evident" sequence of events is not historically accurate. Animal domestication preceded plant domestication and was patently not undertaken to address food insecurity. No one in urgent need of more food and predictable food supplies would take in, feed, nurture, and breed another meat-eating animal in hopes of a payoff in cooperative hunting generations hence. And no one, hoping for a walking larder in the back paddock, would start by trying to domesticate a wolf.[9]

Beginning in the twentieth century, our understanding of domestication grew as we uncovered more facts about its history. And simultaneously, our definition of domestication grew less capacious. In a key conference convened in 1968, Sandor Bökönyi articulated the "essential criterion" of domestication as "the propagation of animals that man keeps in captivity, or more exactly man's breeding of animals under artificial conditions." The precursor of domestication, Bökönyi asserts, is animal keeping, the maintaining of animals without humans being fully responsible for their feeding or their reproduction. I suspect he is right.[10]

This strong emphasis on the genetic changes involved in domestication, and how those might have come about, has characterized the present-day discussion, in which genetic research is often a key line of evidence. Certainly the genetic changes

produced by domestication are critical in distinguishing between a short-term relationship between humans and another
species—taming, cohabitation, or keeping in captivity—and
domestication per se.

However, reproductive control and the genetic consequences
of domestication should not be overemphasized. For example,
a leader in domestication studies, Melinda Zeder of the Smithsonian Institution, offers the following definition:

> Domestication is a sustained multigenerational, mu-
> tualistic relationship in which one organism assumes
> a significant degree of influence over the reproduc-
> tion and care of another organism in order to secure
> a more predictable supply of a resource of interest,
> and through which the partner organism gains ad-
> vantage over individuals that remain outside this
> relationship, thereby benefitting and often increasing
> the fitness of both the domesticator and the target
> domesticate.[11]

I am dissatisfied with Zeder's definition because it is so difficult to assess when an influence over the reproduction and
care of another species becomes "significant," never mind the
impossibility of determining prehistoric intentions. The *why*
of domestication is extremely difficult to deduce if it occurred
in preliterate times and I, for one, do not think intentionality
is an integral component of the domestication process.

However, Zeder rightly emphasizes that, without mutual
benefit, domestication is unlikely to occur at all. Not all commensal relationships proceed to domestication, usually because
of some behavioral trait of the potential domesticate. For example, gazelles and deer have both been tamed repeatedly, yet

neither has ever successfully been domesticated. They both tend to panic in captivity and have been labeled "farouche," a rather charming French word meaning wary, wild, or shy.

When there is mutual benefit, Zeder identifies a commensal pathway to domestication, in which two species come together regularly and develop a relationship of familiarity. The commensal pathway is usually started because the target species is drawn to something about the habitat as modified by humans, such as food scraps, and only later do the two species develop a two-way partnership. Dogs are what Greger Larson and Dorian Fuller call "the archetypal commensal pathway animal." We had something important in common, we humans and future dogs: hunting.[12]

An alternative or second potential pathway to domestication is what Zeder calls a prey or harvest pathway, which is initiated by humans in an attempt to obtain more food or a more reliable source of food by managing the target species. The trouble with this view, as I pointed out earlier, is that people who are short of food are unlikely to embark on saving seed or to spare an animal's offspring from death so that they will mature and produce offspring in the next season.

Zeder also distinguishes a third pathway, calling it a directed pathway because it involves a human intention to domesticate a species. Clearly, the second and third pathways are only likely to occur after the commensal pathway has been successful with some other species. The domestication of the first species ever— of wolves into dogs—cannot have been intentional.

What we have looked at so far are behavioral adaptations to domestication and the archaeological effects in the faunal remains. Can we distinguish different types of domestication? Domestication is generally agreed to be a continuum, a long process, and not an event. As an essential first step, individual members of the wild species need to become less frightened and wary of humans than is normal. Sandor Bökönyi maintains that this is the result of accommodation and adaptation caused by the practice of animal keeping or species management. Familiarity does not breed contempt, as the adage would have it, but lessened reactivity and diminished fear. Scientifically, this is manifested as lower levels of cortisol, adrenaline, and other "fight or flight" hormones, probably in both humans and in potential domesticates. Susan Crockford, an evolutionary biologist at the University of Victoria in Canada, argues that natural variations in hormone levels were a key force in determining which individual animals—in this case, which wolves—were amenable to domestication.[13]

Another way to look at this domestication process and the lessening of reactivity and fear is to see it as an early consequence of starting to develop an effective system of cross-species communication: a sort of behavioral language used by the two species. The best analogy I can think of is a creole or pidgin language, a trading or communication system developed between two peoples by trial and error that is used to bridge two or more different cultures. As the behaviors of each party become more familiar and predictable to the other, the two begin to develop a mutual language of behaviors. This often diminishes fear and aggression. Being together with the other species or culture becomes less stressful and more beneficial as basic ground rules are tacitly negotiated. A commensal relationship, advantageous to both, can be established.

This view relies upon understanding the process of domestication as an agreement between two species to engage together in a particular set of activities. Working with colleagues, Robert Losey, an archaeologist and specialist in Siberian cultures at the University of Alberta, has recently proposed a subtly different perspective on the process of domestication. Domestication is not primarily genetic or morphological, Losey argues; it is a process of repeated, shared activities that lead to *enskilment*. Enskilment is the process of learning how to carry out a task or behavior by engaging in it, over and over, until it and its associated behaviors and objects become familiar. As an example, he evokes

> a person who does not know how to approach, handle, feed, or care for animals [who] will have difficulty controlling their breeding, let alone successfully working with them in tasks such as hauling loads or plowing fields. Animals too become enskilled in being domestic. They come to know our dispositions, feeding procedures, gestures and verbal communications, and even our smells and sounds.[14]

Usually this relationship is assumed to involve the animal living in the same place as and in an environment heavily controlled (and often modified) by humans. The animal, in effect, moves into and adapts to a human world because the benefits of doing so are pleasing and advantageous. This is, as Losey and colleagues observe, only part of the story, for the humans must equally become enskilled at working with wolves or the entire endeavor will end badly. Domestication is, in effect, the process of finding and exploring those key moral values and behaviors that can create a strong bond between two species.[15]

For example, it has often been suggested that a wolf living commensally with a human group may find the warmth, protection from other predators, and extra food to be worth the trouble of adapting to human ways, as long as behaving in ways acceptable and beneficial to the humans is not too difficult. I am somewhat skeptical of the benefit of warmth to wolves, since in the wild they are very well adapted to cold. But the warmth of a cuddling wolf or wolf-dog may constitute a real benefit to a relatively hairless human. Moreover, wolves are highly territorial and aggressive in the defense of their territory and food sources. The wolf's growling when another predator or a strange human approaches might provide important protection, especially for women and children, so that the presence of wolves and humans in cooperative defense may be a substantial benefit to both species. Wolves are more alert to intruders or strangers than most humans and will react more rapidly to perceived threats. Humans have both fire and weapons that might help drive off a challenger.[16]

In hunting, the wolves' superior senses of smell and hearing offer quicker detection of potential prey, their speed makes it possible to pursue and tire prey, and their ferocity and group hunting tactics enable them to surround a prey animal and hold it in place—all traits that seem to improve the success of human hunters substantially. The long-distance weapons wielded by humans might also substantially improve the kill rate and lessen the risk of injury for the wolves.

Juliet Clutton-Brock, another scholar known for her key publications on domestication of mammals, emphasized the fact that dogs also provide companionship and emotional rewards to domesticators. This satisfaction is undoubtedly real, at least to modern pet owners, and has been shown to be an important factor in ameliorating mental and physical disabilities, aging, loneliness, autism, post-traumatic stress disorder, and other modern problems. Animal-assisted therapy is a burgeoning field of mental health treatment. As more people lead urban lives widely separated from family, the emotional companionship afforded by a pet assumes increasing importance.

The emotional benefits of living with another animal have probably been substantial from the outset of domestication.[17] However, it is very difficult to measure or detect emotional satisfaction in the fossil record, so I think we might best look for the tangible results of domestication—how human life or animal life was changed by the formation of this covenant or alliance between species.

Another key issue, raised articulately by Darcy Morey and Rujana Jeger, is that for domestication to be effective you must induce a permanent behavioral and genetic change; in other words, the effort must be sustained. This is the same idea Zeder referred to as "multigenerational."[18]

We can look for such a shift in hunting success by inspecting the time period in the archaeological record during which it is proposed that the first wolf-dogs or Paleolithic dogs were domesticated. What evidence do we have of a shift in hunting success and a growing intimacy and cooperation among humans and dogs? Where does this evidence first appear?

SIX

Where Did the First Dog Come From?

ONE OF THE TRICKIEST aspects of unraveling the domestication of dogs is defining terms and recognizing what the evidence shows. In a real sense, a dog is a domesticated wolf. Another way of putting it is that a dog is a wolf who chooses to live with humans and adapts to an anthropogenic (human-created) habitat. There is no single gene or set of genes that make a canid physically a dog and not a wolf. Nor is there a reliable distinctive marker, such as an extra toe or more teeth, that distinguishes a wolf from a dog. There are tendencies or degrees of anatomical difference that suggest a specimen is from a dog, not a wolf, but these are not simple presence-or-absence traits. Most of the changes that demarcate a dog as opposed to a wolf are behavioral. They are not wholly genetic, but they are surely partly genetic. The origin of dogs has been a hot issue in evolutionary science for more than a decade, and the controversies are still continuing.[1]

Various scholars have set out lists of physical traits that are supposed to indicate domestication, but the physical changes that accompany domestication are difficult to specify. First, so many different types and shapes of animals have been domesticated that drawing overall conclusions is difficult. A few generalities have been proffered. In the case of dogs, one of the classic works was published in 1985 by Stanley J. Olsen, then of Arizona State University, with a chapter by his son John. They specified key skeletal traits of canids that identify a dog, based on a comparison of wolves and modern dogs.[2]

Zooarchaeologists like the Olsens are the specialists who are often expected to analyze an archaeological assemblage of bones and answer the key question of whether the species at a particular site was a domesticated animal or not. Their classic approach to species identification has been to examine the shape of the bones, particularly skulls and teeth. According to the Olsens, traits that distinguish domestic dogs are (1) lop ears rather than erect ones; (2) a shortened and broader snout; (3) smaller and more crowded teeth; (4) a smaller body; (5) a distinct "stop" in the nasal profile; (6) a lower jaw or mandible with a straighter and more vertical ascending ramus (the part of the jaw to which a major jaw muscle is attached); (7) an ascending ramus that is not recurved, as it is in wild canids; (8) a mandible, or lower jaw, with a lower border that is straight, not curving; and (9) archaeological evidence that the canid was living with humans.

How many of these features do we expect to see in the first dog, when domestication began? Not many. Why not? Because animals evolve and change over time. We can almost guarantee

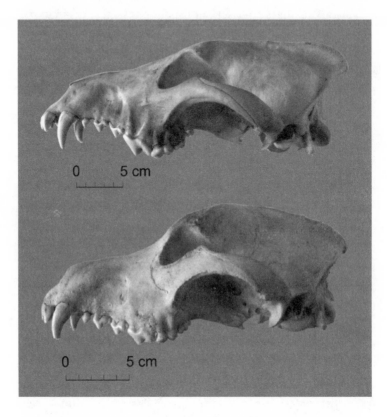

Although the cranium of a wolf (*top*) is similar to that of a large dog (*bottom*), differences can be seen. Note that in this figure the larger wolf cranium is scaled to appear similar in size to the dog cranium to highlight the differences in shape. The profile of the dog shows a strong "stop" in front of the eye sockets, which is typical of domesticated animals. Wolf teeth are usually larger than dog teeth. Because their snouts are shorter, dogs often have crowded cheek teeth and a broader cranium relative to total length.

that the first representative of any species will not look like modern individuals of that species in every detail, whether genetic, or morphological, or even archaeological traits are considered. They must look like their ancestors, at least in large part. So the first dog will look a great deal like a wolf and the first cat will resemble a wildcat.

Further, the anatomical criteria enumerated above are all comparative, in the sense that a shortened snout, smaller body size, more crowded teeth, and so on can be recognized only relative to one's expectations. Sometimes it is easy to look at a modern wolf's anatomy and see how it differs from a dog's, but that ease depends on the breed of dog. Even archaeological criteria are ambiguous because, at least in theory, a canid could live intimately with humans but not in circumstances that resemble those of modern dogs. And, knowing how evolution works, if you look at ancient skeletons, the distinction becomes less and less clear the farther into antiquity you go. Today, domestic dogs are the most variable species known—one that includes a very wide range of shapes, sizes, colors, and dispositions—so the past must also have encompassed a tremendous amount of variability.

For example, we might ask whether the first dog should be smaller in body size than its wolf ancestor. If that earliest dog continued to function and behave much as a wolf did—as a pack hunter chasing down medium-sized to large prey—retaining the hefty size and strength of a wolf would still be advantageous. Besides, not all modern domestic dogs are small or even medium-sized. Some modern breeds of dogs are still as big as a wolf; others are as tiny as a "pocketbook" dog. Being one size or the other does not brand an animal a dog, though size may well reflect the job or ecological niche to which an animal is well suited.

If the first dog still seized prey with its jaws and teeth, as a wolf does, then having a shorter and broader snout might have been of no advantage to the evolving canid. This also means that smaller and more crowded teeth would offer no benefit, unless these traits were a side effect of lower levels of hormones that correlate with aggressiveness. The shorter snout and crowded teeth may have evolved well after the first dog evolved.

The first issue is this: to identify a dog, we cannot use a purely anatomical analysis because modern dogs—the only ones we can be absolutely sure are dogs—have been deliberately bred to emphasize many different anatomical features and many different aspects of performance. Swift-running dogs, such as greyhounds, do not look much like guard dogs, such as Great Pyrenees or German shepherds, and do not look much like scent hounds, such as bloodhounds, either. In the past 200 years, breeding to create different types of dogs has been greatly exaggerated and sped up, producing ever more diversity among modern dogs.

Comparing any ancient dog with only modern dogs is bound to reveal differences. The nature of the comparative sample is crucial. The fundamental way to identify a skull (or a tooth or a jaw) is to compare it with the most likely alternatives, known dogs and known wolves. How many individuals should be in the sample of knowns to show the full variability in the species? Ten? Fifty? Two thousand? More? If we compare a possible dog to known dogs and known wolves, should those dogs and wolves be modern or ancient? After all, evolution has been occurring ever since the first dog appeared; witness the great diversity of modern dogs (Chihuahua to Great Dane, bloodhound to Italian greyhound, shorthaired pointer to dachshund) and modern wolves (Asian, Alaskan, Indian, North African, American, and European).

A recent review of the usefulness of various anatomical measurements of dog and wolf specimens conducted by Luc Janssens and colleagues, using a large set of specimens, revealed that different measures of relative skull, jaw, and tooth size can help "to a limited extent" to distinguish the oldest archaeological dogs from wolves. But as I said before, there is no single anatomical trait that can distinguish a dog from a wolf with certainty. Previously favored criteria were rejected by the Janssens team because of extensive overlap in measurements of dogs and wolves. They concluded, "Upon testing the range of morphometric and morphological variables involved in historical and recent studies for identification of domesticated dog specimens, we found that the majority of these were ineffective at distinguishing between domesticated dogs and wolves." If anatomical features are of little use in distinguishing dogs from wolves, how are scientists to proceed? Some Italian scholars have argued that wolf-dog crosses may show dewclaws, which large dogs tend to have on their hind legs and pure wolves do not. (Nor do pure dingoes show dewclaws.)[3]

Geneticists have played important roles in attempts to trace domestication of various species because domestication involves creating a genetically distinctive species. Two primary strategies are used to trace the origins of dogs (or any other species). The first is the informal rule that the genetic diversity of the species will be greater near the source of its origin because the animal has been in that region longer and has had more time to diversify and evolve. Among other factors, this principle has been used to establish that modern humans arose in sub-Saharan Africa because the genetic diversity of living peoples is greatest there. So where is dog diversity highest? It is a tricky question to answer. Starting in the late 1700s, it became popular in England and English colonies to create new breeds of pedigreed dogs. Registries were set up to define the required traits for a specific breed and to track lineages. That led to deliberate attempts to create a breed of dog (or cat, horse, cow, and so on) that did not look like the others. This meant breeders would try to choose breeding pairs at the extremes—pairs that would feature, say, longer ears, or a fluffier coat, or a shorter face. Thus, human fashion in some parts of the world worked to increase and preserve the diversity among canines. Greater human fascination with novelty would have increased diversity, not time and adaptation. So simply looking for diversity might lead to erroneous conclusions about the place of dog origins.

The second issue is that there must be something in a region of origin for the species of interest to evolve from before the first appearance of the new species. In the case of dogs, the wolf ancestor must have been present in the region of origin. And because we are dealing with domestication, humans must have been present too. We already know that wolves and humans lived in Europe, Asia, and parts of northern Africa about 50,000 years ago. But we can rule out the Americas, Antarctica, and Australia as the homeland of the first dog. People reached those areas too late to domesticate wolves. In some of these regions, there never were any wolves.

The most reliable answers to complex questions such as the origin of the first dog are likely to be found when the genetic evidence, the anatomical evidence, and the archaeological evidence of a human-dog relationship all concur. But there is as yet no such concurrence. The natural movement of animals and the ability of all sorts of canids to crossbreed complicates the story.

By looking at some key studies, we can see how disagreements and confusion arose. No two studies use the same sample of knowns to compare with unknowns, and many analyze only a small portion of the genetic material rather than the whole genome. How much is enough? It is as if you were trying to document the average height of a species using tape measures that vary in the size of a foot or inch.

In 1997, a large-scale study of the mitochondrial DNA (mtDNA) of various wolves and modern dogs was carried out by Bob Wayne's genetics laboratory at the University of California, Los Angeles. They investigated a segment (261 base pairs in the control region that turns genes on and off during development) in the mtDNA sequences of 162 wolves from twenty-seven localities worldwide and of 140 domestic dogs representing sixty-seven breeds. This study revealed some key facts. The samples of dog and wolf mtDNA showed a maximum of only twelve amino acid substitutions, whereas dog mtDNA differed from that of other wild canids—such as coyotes or jackals—by at least twenty substitutions. This is strong evidence that gray wolves were the ancestors of dogs—a conclusion now rarely challenged. The team found twenty-seven different haplotypes (or genetic lineages inherited as a unit) among wolves and twenty-six among dogs. Only one haplotype was shared by both dogs and wolves. Some haplotypes (four in wolves and four in dogs) were very widespread geographically, but most appeared to be common within particular geographic regions.[4]

This study also showed that dogs fell into four large genetic groupings, or clades, each of which shares a common genetic heritage. Because many dog breeds exhibit haplotypes or genetic lineages from several clades, clades are not clear breed markers. Wayne's team referred to these clades by Roman numerals I–IV, but the terminology shifted in later studies. Some 73 percent of the dog breeds analyzed fell into Clade I, now

generally called Clade A. This clade included so-called primitive or basal breeds, such as the Australian dingo (which might
be a separate species, not a type of dog or wolf), the African
basenji, the Chinese chow chow, and the greyhound, which is
believed to have originated in ancient Egypt, Persia, or Greece.
In other words, some dogs in Clade A are evolutionarily primitive but they came from scattered geographic regions. Wolves
also clustered in Clade A.

Two Scandinavian dog breeds, elkhound and jämthund, fell
into Clade II (now called Clade D). Clade D haplotypes are
closely related to two wolf haplotypes found in Italy, France,
Romania, and Greece.

Clade III, now known as Clade C, contained only three dog
haplotypes found in several breeds, including the German
shepherd, Siberian husky, and Mexican hairless. These haplotypes are not clustered geographically and may have evolved
independently of each other.

Clade IV—now called Clade B—also had only three haplotypes. These were identical or very similar to wolf haplotypes
found in Romania and western Russia. It is speculated that
these haplotypes are probably the result of recent hybridization between dogs and wolves.

Some of the overall results from Wayne's study were later
challenged by a group led by Peter Savolainen, who had been
a postdoctoral fellow in Wayne's lab and a coauthor on the
previous paper. Savolainen was then working with Chinese
and Swedish collaborators. In 2002, the cover of the presti-
gious magazine *Science* featured an image of a yellow Lab-
rador retriever, illustrating an article on the subject of the domestication of dogs by Savolainen's team.[5]

The team analyzed a longer portion (582 base pairs) of
mtDNA in thirty-eight Eurasian wolves and 654 domestic

dogs from Europe, Asia, Africa, and Arctic America. They started with an assumption that dogs originated earlier than about 14,000 years ago, relying on the dating of one fossil dog jaw from Kesserloch, Switzerland, and another from Bonn-Oberkassel, Germany. The Kesserloch dog is generally accepted as an early dog, partly because it is smaller than a contemporary wolf and partly because the teeth are crowded in its small jaw, both criteria offered by the Olsens. The Bonn-Oberkassel fossil was covered in ochre and buried both with and like two humans, one male and one female. Such ritual treatment signals an intimate relationship between dog and human—a wild canid that served as food or was merely a defeated competitor is not likely to be carefully buried. Burial, in short, is an indicator of the status or importance of the buried individual, whether it is a human or an animal given humanlike status. Deliberate burial is just about the gold standard in terms of evidence that an animal was domesticated.

Burial has been used as a criterion since a 1978 paper on the archaeological site of Ein Mallaha in Israel. The site had belonged to members of the Natufian culture, which created some of the earliest permanent settlements, with semisubterranean houses with stone walls. The site was radiocarbon dated to between 11,740±740 and 11,310±570 years before present (redating with modern techniques would probably revise this date to be older). The Natufians at Ein Mallaha are regarded as just preagricultural.[6]

In the burial, the woman's hand rests on a puppy, implying that it was her pet. That is strong evidence of intimacy between canids and humans, often called a "special relationship" and a hallmark of domestication. The Ein Mallaha burial highlights crucial evidence that dogs cohabited with humans and were treated much like humans after death. As the archaeologists involved, Simon Davis and François Valla, observed, "The puppy, unique among Natufian burials, offers proof that an affectionate rather than a gastronomic relationship existed between it and the person with whom it was buried."[7]

In their work, Savolainen's team also relied on the assumption that the dog clade with the greatest genetic diversity would represent the most ancient one. Like the samples analyzed by Wayne's laboratory, Savolainen's samples fell into four clades, and a few apparently represented a new fifth clade. Confusingly, Savolainen's group coined a new terminology for the canid clades, calling them A–E, with Clades A–D basically the same as the Wayne team's Clades I–IV. The largest clade represented in Savolainen's study was the most numerous and showed the broadest diversity, including haplotypes from Chinese and Mongolian wolves along with those from dogs. This clade had a larger total number of haplotypes (forty-four) and a larger number of haplotypes not found elsewhere (thirty). Logically, this diversity would identify the oldest dogs. Also, the East Asian samples showed a greater percentage of unique haplotypes than each of the other regional groups (Southwest Asia, Europe, the Americas, Siberia, India, and Africa). On this basis, Savolainen's team concluded that East Asia was probably where dogs were initially domesticated. In their estimates, which were based on the number of mutations, the origin of Clade A may have occurred about 40,000 years ago, or, if Clades A, B, and C were founded at the same time, about 15,000 years ago.

Wayne's group, among others, noted that this conclusion did not match the archaeological evidence well because the oldest fossils of domesticated dogs come from Europe or the Middle East, not Southeast Asia or China. If dogs were first domesticated in Asia, where were the remains of all of those dogs?

The archaeological and fossil record from East Asia shows evidence that the first dogs were not living there with humans until about 9,000 years ago—not 40,000 or even 15,000 years ago. The oldest Asian dogs are represented by more than 110 individual dogs that come from burials attributed to the Jōmon culture in Japan, and those burials are only about 9,000 years old. Unfortunately, the Savolainen study did not sample any of these earliest East Asian dogs because of the difficulty in extracting ancient mtDNA from fossils at the time. (Better techniques have since been developed.) The oldest dog remains from China itself are even younger, only about 7,500 years old, which also challenges the conclusion that the first dog evolved in China. There is an uncomfortably large chronological gap between the earliest buried Asian dogs and the well-accepted early dogs at Bonn-Oberkassel, Kesserloch, and Ein Mallaha (14,000 to 12,000 years ago). Could domestic dogs have lived in East Asia for millennia but remained invisible in the fossil record? Maybe the tradition of burying dogs as if they were people simply did not exist in Asia—but why were more recent dogs from Jōmon and the earliest Chinese dogs buried?

Recall that it has been suggested that domestication might be recognized by skeletal hallmarks, including having a smaller body size than the wild ancestors and, in dogs, shorter faces with smaller muzzles, which produces a distinct "stop" in the profile of the skull just at or below the orbits. Shortening the face often means crowding the teeth, too. But these are problematic criteria for many reasons.

In the case of the Ein Mallaha canid, this way of distinguishing dogs from wolves was tried; but because the puppy in the burial has milk teeth, Davis and Valla did not know how to evaluate the teeth. In adult dogs, the first lower molar and the tooth next to it, a premolar, often overlap. Whether or not the degree of overlap remains constant throughout life is unclear. As Davis and Valla remark drily, "dental overlap criteria are not efficacious in dog-wolf separation." (This remark apparently outraged some, such as the Olsens, who used this characteristic often.) Nor, of course, would body size be a reliable criterion. Wolves are usually larger than dogs; a cub, or a puppy, is certainly going to be smaller than an adult. The likelihood is enormous that the first dog would have looked almost exactly like a wolf. Whatever the fossil record shows (or doesn't show), it is true that some attributes of domestication are almost impossible to discern from fossil remains, such as lop ears rather than erect ears, or piebald or mottled coats, or behaviors such as being less reactive than a wolf to novelties, including people. If a wolf cub is exposed to benevolent humans early enough—the period for socialization starts at about two weeks after birth and continues for a month—then the cub will be much less aggressive and less fearful of humans for the rest of its life than a wolf cub without such exposure. However, the timing of this window for socialization is two weeks earlier in wolf cubs than in puppies, occurring when the wolf cubs are still deaf and blind. In stark contrast, puppies can see, hear, and smell during the socialization period. Thus even wolf cubs that are exposed to humans very early in life learn much less about people than dog puppies do.[8]

How can we possibly spot these behavioral changes in the fossil and archaeological record? We could look for evidence of a change in lifestyle, a more common association between humans and canids, and the development of a special relationship between wolf-dogs and humans that might lead to a tendency to treat dogs as if they were people.

The claim for the earliest dogs is based on a morphological analysis of fossil canid skulls in Europe published in 2009. Belgian paleontologist Mietje Germonpré led a team who used sophisticated statistical techniques to identify a group of dog-like fossils ranging in age from 36,000 years ago at Goyet Cave, Belgium, to 26,000 years ago at Předmostí in the Czech Republic. The team identified another dog cranium at about 15,000 years ago in Eliseeivichi, Russia, and two more at 17,000 and 14,000 years ago from Mezhirich and Mezin in Ukraine. Yet another very early "incipient dog" has been identified at Razboinichya by Nicholai Ovodov, Susan Crockford, and colleagues, who examined a 33,000-year-old canid from Siberia using a different comparative database but similar techniques. Though "incipient dog" is not a well-defined term, the group analyzing remains from Razboinichya seems to mean a canid early in the process of domestication.[9]

A team from Bob Wayne's group, led by Olaf Thalmann, subsequently analyzed part of the mtDNA control region from the Razboinichya canid, showing that its DNA grouped with wolves and two Scandinavian dog breeds in Clade D. They also found that the mtDNA of several "dogs" from Goyet Cave was very basal, positioning these remains in a more primitive or wolflike sister group to all modern dogs and wolves sampled. However, there is no DNA marker that is definitive for dogs and not wolves, meaning the DNA groupings are ambiguous. Both the Goyet "dog" identified by Germonpré's team and the Razboinichya "incipient dog" identified by Ovodov and Crockford may have been early failed attempts at the domestication of wolves. By failed attempts, perhaps scholars mean attempted domestications of animals that, in the end, were not well suited to traveling and living with humans.[10]

SEVEN

Interwoven Stories

THE PRIMARY TRUTH emerging from all these studies is that dogs lived and traveled with humans just about everywhere that humans went. Dogs, like people, are capable of playing so many different roles that they are useful to humans in a very wide range of habitats and lifestyles.

If dogs are the ultimate—or the primordial—human companion, we might well ask when and where they first accompanied us. What we have is several interwoven stories of how humans evolved and moved around the world, under different circumstances and in different ecosystems. Some of the biggest questions in paleoanthropology and prehistory today focus on human migrations and adaptations, particularly the initial spread of modern humans out of Africa and the demise of archaic humans such as Neanderthals and Denisovans.

I have argued that a major factor contributing to the success of anatomically modern humans over archaic humans in

Eurasia may be the unprecedented alliance between the modern humans and wolves that were being domesticated into dogs. A broader but equally important question is whether the domestication of animals—the unusual cross-species cooperation that started with the domestication of the dog—was equally beneficial elsewhere in the invasive success of modern humans. Is the story the same in each major migration?

Sadly, we have probably not reconstructed the history of human migrations correctly by focusing so strongly on the European story (most of the scientists working on these issues have been European). I will review that story briefly here before looking to see what we might have missed.

Probably the biggest surprise for our ancestors as they expanded their territory out of Africa into Europe was encountering Neanderthals, who had been living and evolving in central Europe for hundreds of thousands of years. Neanderthals were both unnervingly similar to and frighteningly different from the African migrants. Consider what you would think if, in your neighborhood grocery store, you unexpectedly encountered people of a very different body type, with a skin color you had never seen before, who had perhaps painted that skin and were wearing odd clothing or hairstyles. I think they might seem alarming and possibly dangerous to many Americans or Europeans. And yet, ethnographers who have lived in parts of the world where communication with outsiders is rare point out that a new person, even a complete stranger, is also a possible source of valuable information. A traveler carries knowledge. Long before any modern means of communication—newspapers, radios, television, cell phones—travelers spread the news in remote areas.

I learned about the "bush telegraph" firsthand in Liberia in 1971. Each traveler or newcomer passed along whatever important information they had to anyone they might meet, wherever they went. Some people could speak four or five different languages, as well as English, even if they could not read and write. As a medium for communication, the bush telegraph functioned remarkably well. I spent some weeks at a remote, up-country missionary hospital, and I was there on the day when the then-president of Liberia, William Tubman, died of old age. To my great surprise, the patients in the up-country hospital knew of Tubman's death before I did, despite having no newspapers, radios, or televisions (and of course no internet).

A comparable event had not happened in most Liberians' lifetimes. Even at private Western-style schools in the capital, the succession of the presidency was not taught, so no one knew what would happen. It was a potentially dangerous time; even minor government officials were reluctant to do anything for fear it had suddenly become the wrong thing to do.

One afternoon, two men in masks, skins, and raffia costumes danced around the compound where everyone lived to the beat of a drum. Openly watching the ceremony was forbidden, though I peeked out the window once or twice. We could hear drums from other villages too. Then many of the staff of the hospital—well-educated people who were nurses, aides, technicians, pharmacists, and clerks—disappeared and took off for their natal villages, "going bush" for fear that their education and status would make others jealous and bring reprisals from members of other ethnic groups. It would be safer not to be noticed and not to be different. Rumors spread uncontrollably—any clue as to the future was of tremendous importance. So my assumptions about information sharing were overturned; modern communications were great if you had them, but people knew that it was crucial to share information and found effective ways to spread knowledge.

Neanderthals may have experienced apprehension when faced with early modern humans and felt the same urgent need to gain information about them. Neanderthals were the archaic humans most similar to us, and they shared much of our lifestyle and behavior, but not all of it. They chased the same prey and sometimes lived close by modern humans—sometimes in the very same caves and rock shelters, though probably not at the same time. This deep similarity means that Neanderthals would have felt the increase in competitive pressure caused by modern humans' arrival in Europe more acutely than any other species. As I have suggested elsewhere, the increased competition from anatomically modern humans provides a special insight into the fairly rapid extinction of Neanderthals by about 40,000 years ago. Apparently, modern humans outcompeted archaic humans in 10,000 years or less.

Factors other than increased competition were also important in the extinction of Neanderthals. Neanderthals were probably always thin on the ground, but the ecological crisis created by the appearance of a new top predator made it particularly hard for Neanderthals to survive things such as climate change. Modern humans' confrontation with Neanderthals and the Ice Age fauna of Europe is the first thread of the story.

As humans moved into the Middle East and Eurasia, they encountered a different group of animals than those they had hunted in Africa: the Ice Age fauna of the continent. These animal species were new to modern humans and yet not entirely novel. In some ways, the basic Ice Age fauna of Europe broadly resembled the African fauna. For example, there were many prey species in Eurasia that would have seemed somewhat familiar. The woolly mammoth of Eurasia was in many ways like the African elephant; the woolly rhino of Europe was akin to the African rhino; and the primitive horses of Eurasia had much in common with the zebras, quaggas, and wild asses of Africa. Though the numerous antelopes and bovids (cattle) of Africa were not all represented by Eurasian forms, the cervids (deer) and bovids of Europe in part resembled the antelopes and bovids of Africa. In terms of predators, both ecosystems featured many large and ferocious felines (e.g., lions, cave lions, leopards, cave leopards, cheetahs) and medium- to large-sized canines (wolves of several species, Cape Hunting dogs, jackals, small foxes). As in Africa, the predator guild of Ice Age Europe was packed with well-adapted species with which early modern humans competed for prey and other resources. But there was enough prey, of diverse sizes and habits, to support a wide range of predators.

Early modern humans became the most important rivals of the archaic humans who had been living as hunter-gatherers in Europe for hundreds of thousands of years. Some of the newcomers' knowledge about ancient African species was applicable in Eurasia. However, the European climate was colder and wetter than that of ancestral Africa, requiring new adaptations.

In some ways, the most significant new mammal early modern humans encountered when they entered northern Europe and Asia was the gray wolf—another large, ferocious, and formidable competitor for the resources that were needed to stay alive. Like modern humans, and like Neanderthals, wolves lived in family groups, hunted in groups, and raised their young cooperatively in communal dens or safe places. Wolves' killing weapons—big teeth and strong jaws combined with running speed, excellent scent perception, and stamina—were part of their bodies, not manufactured items such as spears or hand-held cutting tools. And at the same time that Neanderthals were hurtling toward extinction, the first hints of dog domestication appear in the record. These are the skeletal remains identified by Germonpré's team that might be called proto-dogs, wolf-dogs, Paleolithic dogs, or simply weird wolves. How to describe these animals is controversial; even the researchers involved cannot agree on how to classify them.

As noted earlier, the team led by Mietje Germonpré has statistically identified more than forty fossil specimens from Europe that form a coherent group of canids more similar to primitive dogs than to wolves in the shape of their crania and jaws. But they are not modern dogs, and the wolves in the same Ice Age ecosystem from which the wolf-dogs are descended were not identical to modern wolves. What Germonpré's team has shown is that these specimens are distinct from wolves of the same prehistoric age in terms of their proportions and morphology (anatomy), their mtDNA, and their diet. Their mtDNA does not match any yet found in either modern dogs or wolves. So, were these animals dogs?

On current evidence, we cannot link these ancient canids directly to any living dog; they are not certain ancestors. Were they simply a strange group of wolves? They were demonstrably unusual. Were they not yet dogs (the domestic species) but no longer wolves? Do they show us what a first attempt at domestication looked like? Possibly. My intuition is to say yes. These wolf-dogs have been found so far in archaeological sites made by modern humans, not in Neanderthal sites. And nearly all of those sites represent something novel at the time, a different way of living and hunting that may have been made possible by an association between humans and these wolf-dogs.

The history of interbreeding and extinctions among modern humans, Neanderthals, and Denisovans is a deeply tangled enigma at present. What we know is that there was more variability among modern or near-modern humans than we expected and that different species—if that is what the hominins were—sometimes interbred.

But only one of the early human groups we have tracked, the anatomically modern humans that reached central Europe, were closely associated with dogs. In my previous book, *The Invaders*, I hypothesized that modern humans outcompeted Neanderthals under the extreme conditions of Ice Age Europe because of a peculiar alliance between modern humans and wolf-dogs. Of all of the members of the Ice Age predatory guild in Eurasia, the only two that survived in the long run were wolves and modern humans. Dogs—once wolves—became the primary companions and collaborators of human hunters, the first animal to work with humans in a systematic way. Did it happen between 50,000 and 40,000 years ago? Not everyone agrees, but the hypothesis still looks plausible scientifically. Turning a wolf into a dog is not something that was done consciously or quickly, but we know it happened.[1]

It may seem paradoxical that such a powerful predator as the wolf evolved into our first companion, rather than remaining our predator and competitor. Wolves are still greatly feared by humans. In Africa, our evolutionary homeland, canids were much smaller and less dangerous competitors than the wolves of Pleistocene Eurasia. African jackals are much like small wolves or coyotes in the Americas. (It has recently been proven that some of the North African canids once taken to be golden jackals are in fact a previously unrecognized species of wolf, now known as the golden wolf.) Africa also had Simien or Ethiopian wolves, as well as African or Cape hunting dogs—leggy, lightly built, pack-hunting canids with blotchy coats—who are very effective hunters, but neither as large nor as successful in bringing down large prey as wolves. Though hunting dogs are very able predators, the extra size and strength of wolves is an impressive advantage.

When wolves and humans met up in Eurasia, something mysterious occurred. For a long time, paleoanthropologists concentrated on the fact that many early humans moved from Africa northward into Europe and then still further northward and eastward into eastern Europe, Siberia, and eastern Asia, and then on to the Americas, possibly taking refuge from harsh climate changes in the now-drowned land mass known as Beringia, which connected eastern Asia to western North America. Wolf-dogs apparently moved with these people or evolved from local wolves while both canids and humans were probably confined to Beringia. These new canids evolved into an animal that was specialized for a niche that included humans. This fact is shown by the close genetic resemblance between the distinctive dogs of the Americas and those of Siberia. Even those who doubt the evidence for early wolf-dogs in Europe agree that having dogs probably made a significant difference to our American ancestors.

Around 15,000 or 14,000 years ago, humans were living with canids that everyone agrees looked like dogs, not wolves or wolf-dogs, and were treated as dogs. The northward-moving people left large dog cemeteries in Siberia in the Cis-Baikal and smaller ones in the Trans-Baikal region to the east. Their dogs were buried deliberately and often with ochre, grave goods, or items of jewelry such as bead necklaces. But we don't know if those first obvious dogs were the descendants of the wolf-dogs or simply another group of quasi-domesticated canids. Possibly all the early European wolf-dogs died out and were replaced by dogs from farther east in Asia, like those from Siberia. This is a scenario that is commonly accepted on the basis of a massive study of eastern and western dogs by Laurent Frantz. The story is confusing, in part because it is incomplete.[2]

What's missing?

EIGHT

The Missing Dogs

THE COMMON European migration
story is that modern humans came from Africa, with an emphasis on one of the last great human territorial expansions
into Eurasia. Unfortunately, this means ignoring the first territorial expansion. Before this last migration into Eurasia,
some people had already left the main group of early modern
humans whose descendants eventually ended up in Europe,
Asia, and later the Americas.

These people were a group of early modern humans who
split off from their kin in the Levant or southern Eurasia and
eventually ended up in Greater Australia, a supercontinent
comprising mainland Australia, New Guinea, Tasmania, and
many other areas that are now islands (or are completely submerged). How did people get to Greater Australia? Because sea
levels were lower than they are now at various times in prehistory, these early modern humans probably took a coastal
route southward and eastward, arriving in Greater Australia

about 70,000–65,000 years ago. They were travelers. This timing suggests they separated from the early humans who moved into Europe soon after modern humans first left Africa. We can call them the First Australians or the ancestors of modern Aboriginal peoples. They were unquestionably modern humans, but they were dogless.

We do not know much about the precise route these people took to end up in Australia—not that they were intentionally going to Australia, because no one knew the continent existed. Unfortunately, the fossil and archaeological evidence from southern Europe and coastal Asia is genuinely paltry between about 70,000 and 50,000 years ago. It has been estimated that Greater Australia lost about 70 percent of its land mass because of rising sea levels. No wonder it is hard to find evidence! We do know that the last part of the First Australians' passage to Greater Australia must have involved several open-water crossings as they moved from mainland Asia (a region known as Sunda) across the island region known as Wallacea, then across water to the myriad islands of southern Asia and on, eventually, to Greater Australia, or Sahul.

The demands of this series of water crossings had probably prevented still earlier hominins, such as *Homo erectus*, from getting to Australia, though the skeletal remains of *Homo erectus* are preserved in what is now Indonesia and China. This archaic hominin was in the right place to move into Greater Australia, but did not have the right skills and information. Some hominins reached the Philippines and some landed on the island of Flores in present-day Indonesia, where they seem to have undergone a phenomenon known as "island dwarfing." Because resources such as game, edible plants, fresh water, and secure shelter are inevitably limited on islands, many island species either become physically smaller—they are dwarfed—

or much larger. So, for example, ancient Flores had little ele-phants (dwarf stegodons) and giant rats. The same phenom-enon of competition for resources may account for the evolution of very small hominins called *Homo floresiensis* on Flores and *Homo luzonensis* in the Philippines. Both of these species are very small in stature, maybe a little over one meter in height as adults, but neither closely resembles hominins from elsewhere. Known *Homo floresiensis* bones were fossil-ized from about 100,000 years ago to about 60,000 years ago and come from about a dozen individuals. This species was not simply one odd person, but a population that persisted over time. In the Philippines, *Homo luzonensis* is dated to about 50,000 years ago but is represented by remains of only three individuals and a few bones.[1]

Apparently these diminutive species did not possess mari-time skills, though they did make stone tools and succeeded in hunting animals on their islands. Island dwarfing is unlikely to have occurred if it had been easy to get off those islands. There are also no archaeological remains that suggest mari-time skills, regular fishing, or consumption of crustaceans or shellfish. So, although archaic species of humans did not make it to Australia, early modern humans did. How and why?

Investigations of the different routes began with Joseph Birdsell in the 1970s and have continued since. Various routes have been suggested to see which sea voyages would have been shorter, and therefore easier, given the currents, tides, and various configurations of land masses. None of the voyages were truly easy. Stegodons, a type of elephant, were the only other medium- to large-bodied terrestrial mammal ever to cross from Sunda (Asia) through Wallacea into the islands west of Great Australia in numbers sufficient to establish a viable population in the new area without human assistance. At least eight and perhaps as many as seventeen ocean crossings, taking altogether four to seven days and maybe more, were needed for humans to inhabit Australia once they had boats and some maritime skills. Included among those vital skills would be fiber art, for making rope or twine, baskets, and nets; vessels suitable for holding fresh water; and a secure means of holding multipart tools together. If people came from Timor to northwestern Australia, for example, the open-water voyage may have been up to ninety kilometers long.[2]

When sea levels were lower and the land masses larger, there were only a few places along this probable coastal route from which the opposite shoreline would have been visible. The distance from various potential launch points to Greater Australia was much too far to swim, totaling at least seventy kilometers in various voyages. The water crossings alone testify that the First Australians had skills in boat-building, coastal living, and open sea navigation. William Noble and Iain Davidson were the first to realize that boat-building and maritime adaptations demonstrated modern cognition in three aspects: increased information flow, substantial planning depth, and the capacity for conceptualization. They argue that the behavioral changes signified by reaching Greater Australia showed that these modern humans had language.[3]

Curtis Marean, an archaeologist at Arizona State University, has been thinking about the development of coastal adaptations among early peoples for years. What he has discovered applies well to the people who became the First Australians, though his primary research has been carried out at early modern human sites such as Pinnacle Point in South Africa. There, remains excavated from a cave (known as PP13) on the shoreline showed that early humans had begun to exploit shellfish regularly about 160,000 years ago. They did not abandon terrestrial hunting but simply enlarged their diet to include these new and valuable marine resources. They made fires, used pigments frequently (almost 400 pieces of ochre had been excavated from PP13 by 2010, all of which showed grinding wear or other signs of use), and created small, finely made stone blades that could be hafted in projectile weapons such as spears or arrows. Sometimes the inhabitants of the cave even heated silcrete, a type of stone, because doing so improved the properties of the material for making tools. The regular use of shellfish and the heat-treatment of silcrete are the earliest known records of such behavior. These were incredibly important discoveries that opened up a new ecological niche for humans in terms of access to new foods and materials for making tools.[4]

Reconstructing the ancient climate in South Africa suggested to Marean's team that the time sampled by the cave remains, starting at about 195,000 and continuing until 125,000 years ago, was an especially long, cold, and difficult glacial period in the world. On a worldwide scale, this period is known as marine isotope stage 6. As a general rule, such a harsh glacial period is marked in sub-Saharan Africa by increased aridity, which would expand deserts and dry lands. Thus the animals and plants that had been staple foods earlier in human history would become scarcer. The discovery of abundant coastal foods—shellfish, fish, and the occasional marine mammal—was a boon to survival. Coastal foods are relatively easy to obtain, and they are rich in protein, calories, and fats. Beach stones for making tools and outcrops of pigments were also important and available. The remains of these newly exploited resources are common in the archaeological record at Pinnacle Point, which may have been a sort of refuge for early humans under difficult climatic conditions.

Marean hypothesizes that using marine resources may have spurred the development of one of the puzzling complexities of the human species. Humans have a tendency to work cooperatively in groups, coupled with a tendency to defend stable and key resources against outsiders. Because each clam or mussel is a fairly small packet of food, working cooperatively and inventing ways to carry numbers of shellfish efficiently would have been important. And since shellfish beds cannot be moved rapidly, an area with concentrated shellfish beds would be valuable enough to be worth defending from other human groups.

In a similar way, marine resources may have been vital to the human populations that took the coastal route to Australia. By moving along the southern coast of Asia, they could continue to exploit coastal resources with which they were familiar. The climate would have been somewhat milder than farther north, too. In time, these coastal dwellers apparently began to experiment with nets, fish traps, and simple devices that functioned as rafts or boats. Put simply, as Marean discovered in his studies in South Africa, the people who became the First Australians learned the language of the ocean and shore, and broadened their diet as well as their ability to move. What is especially fascinating is that some of the innovations found in Pinnacle Point—exploitation of coastal resources, use of ochre, use of heated silcrete, and the development of small blades hafted onto a larger piece, like teeth in a saw—have also been found in the earliest archaeological site in Australia.

As these future Australians continued to learn, their ocean voyages grew longer, their seamanship became more adept, and their ability to move from one island to another with little effort improved. In the end—or the beginning, depending on how you look at it—these early modern humans stumbled onto an island so big they could not see its full extent. Greater Australia (Sahul) must have appeared endless. And it was full of new animals and birds so naïve that most were astonishingly easy to capture and kill. Madjedbebe, the earliest archaeological site now known in Australia, shows that humans arrived in Sahul by about 65,000 years ago, at least 15,000 years earlier than other modern humans reached central Europe. Even scholars skeptical of that date are fairly comfortable with a date of 50,000 years ago for landfall in Greater Australia. The thread of human evolutionary and migratory studies that initially concentrated on developments in Europe was turned upside down by the discovery of such early sites in Australia. Australia was first.[5]

Of course, modern humans had no way of knowing they had entered a new continent, but they must have realized that there were plants, fish, animals, and landscapes that were very different from those they had known elsewhere. Sahul was filled with mammals, many of which were decidedly peculiar, such as kangaroos and possums. Though kangaroos' heads are eerily similar in shape to those of antelope or deer, none of the herbivores our ancestors knew from Africa hopped on two legs, stood upright using a large tail for balance, and raised their newborns in pouches. With the exception of a few bats, rodents, and odd, egg-laying, indigenous mammals such as echidnas and platypuses (both classified as monotremes), all of the medium to large mammals in Greater Australia were marsupials with an entirely different reproductive system from that of the placental mammals of Africa and Europe. Among the consequences of the reproductive system of marsupials is the fact that their offspring are all born in a very premature—almost fetal—state of development. None of the offspring get up and run or jump minutes after birth, as African antelopes or Eurasian deer do. Marsupials are very slow developers and live a long time within their mothers' pouches.

By accident or design, taking the coastal route across Asia meant that future Australians less often encountered the northern predators that were similar to the felids (cats), canids (dogs), hyenids (hyenas), and ursids (bears) of temperate and Ice Age Europe. Taking this route also avoided the extremely cold temperatures and the special problems of survival in central and northern Eurasia. However, the future Australians who took the coastal route as they moved farther east left very few remains behind in India, southern China, Island Southeast Asia, or what is now Indonesia.

As best it can be reconstructed, the southern coastal route would have offered some important opportunities. Once people learned how to fish and, especially, how to collect shellfish, they had new and valuable food resources that were predictable, rich, and dense in calories. Though fish had to be speared, netted, caught on hooks, or even captured by hand using techniques that were invented during the journey, shellfish were easy to collect and did not have to be pursued the way a kangaroo or wallaby did.

Marean defined a *coastal adaptation* as "a lifeway in which coastal materials play a significant and central role in nutritional, technological, and social aspects of a given community." People with still more advanced technology develop a *maritime adaptation,* in Marean's terms, which "includes the use of open-ocean boats to assist in the collection of nutritional and raw material resources."[6]

Marean's definitions are applicable to the settling of Australia particularly as concerns Asitau Kuru (once called Jerimalai) and Lene Hara, important sites on the island nation of Timor-Leste off the western coast of Australia. The different stratigraphic layers at Asitau Kuru are tightly dated, using state-of-the-art techniques. The oldest layer is between 38,000 and 42,000 years before present, and the youngest artifact-bearing layer is between about 5,000 and 4,600 years ago. In addition to large numbers of chert artifacts, there are bone points, fishhooks, and beads made of shell, as well as faunal remains that persist throughout the deposits. The faunal remains indicate clearly that the people who lived at those sites were adept exploiters of marine resources. The researchers recovered more than 38,000 fish bones representing at least 796 individual fish. Almost half of those bones (by weight) are from pelagic species, such as tuna; the rest are from inshore fish. In other layers, marine turtles dominate the remains. The remains of seagoing fish demonstrate that these people made and used boats of some sort. Asitau Kuru has yielded only two fishhooks, both made of shell and both from later periods in the site. The oldest fishhook (the oldest known fishhook in the world) can be dated to between 23,000 and 16,000 years ago. The younger fishhook is more complete. Both were made for single-bait line fishing. The people at Asitau Kuru also left bone points, made from the spines of larger fish, that may have been part of a composite tool for fishing, such as a fish spear or a multiple-hook device for trawling. Marine resources were not exploited casually or occasionally but systematically, using boats and special technology. These were repeated endeavors carried out for at least 38,000 years. The site also holds small quantities of the bones of rodents, bats, birds, and various terrestrial reptiles such as pythons and monitor lizards.[7]

Ongoing excavations at Asitau Kuru have yielded five remarkable objects made of worked and polished *Nautilis pompilius* shells, some of which had been colored with red ochre. The shell fragments had been polished to remove the outer

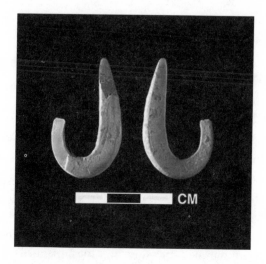

This fishhook made of shell supports the idea that early humans in Island Southeast Asia relied on maritime resources. The fishhook is the younger of two excavated in a cave on Timor now called Asitau Kuru (formerly Jerimalai); one was dated to 9,741±60 years before present, with a layer immediately below the fishhook yielding a shell dated to 10,050±80 years before present, making it the oldest evidence of such technology in Island Southeast Asia. Fish remains include pelagic species such as tuna, indicating that the boats were taken out to open water. Each block in the scale is 1 cm.

white layer of shell and expose the nacreous mother-of-pearl inner layers. (Experiments with a modern beach-collected *Nautilus* shell showed archaeologist Sue O'Connor and her team that locally available fine sandstone could be readily used to polish away the outer layer.) Although the inner layer of the shell is naturally nacreous, the outer layer of the worked fragments had been abraded with hematite or fine pumice to produce a higher gloss. In all, the team recovered a complete polished and pierced pendant made of shell and several pieces of a broken pendant that fit together. Four of the *Nautilis* objects were colored with ochre, which does not occur in the sediment from which they were excavated. In addition to the shells' natural reddish-brown and white coloration, the use of ochre to create a red and white decorative pattern suggests that these patterns were symbolic and important.

O'Connor and her colleagues are cautious in their interpretation of these remains. In Island Southeast Asia, *Nautilus* have occasionally been eaten but are more often valued for the large shell and its distinctive coloring. The team argues that the rare but persistent use of *Nautilus* shells in the manufacture of decorative items indicates that this indigenous species was a special symbol or signifier of the people who created the Asitau Kuru site. The coast provided not only food but also an opportunity to create symbolic objects unique to a particular group. Similarly, beads created from different shell species much earlier in time in Africa or Europe were long-lasting symbols or identifiers of a group.

Considering the Asitau Kuru finds as a whole, we can see that early humans arrived in Southeast Asia close to Sahul by at least 42,000 years ago, and probably earlier. These humans developed a clear maritime adaptation, using marine resources from close to the shore as a regular source of food as well as more distant oceanic habitats reached by boat. In addition, they used particular, endemic marine resources to create decorative objects that may have symbolized their membership in a particular local or regional kin group or tribe. Living *Nautilis* frequent deep water and might be difficult to catch, but the shells themselves are washed ashore naturally after the mollusk dies.[8]

This level of sophistication and exploitation of marine resources is exactly what might be expected from the First Australians. Though the earliest archaeological site on Greater Australia is now dated to about 65,000 years ago, the presence of a maritime people as close to the mainland as Timor is completely congruent with a human migration from Sunda to Sahul.

As people began to rely on the sea and its resources more, they learned how to construct some kind of boat or watercraft and rope, twine, and maybe sails. They adapted, innovated, discovered, and ended up in a brave new world that was Greater Australia. Their evolution, their migration, and their skills form the second thread of our story. Because much of the migration itself remains mysterious, we know as much about the *settling* of Australia as we do about *getting to* Australia.

One of the few undisputed facts about the first humans' getting to Sahul is that it must have involved a series of open-water voyages from Asia. The demands of this water crossing may well account for the limited number of occasions upon which new species entered the continent. The distance to Australia from various points in Island Southeast Asia varied historically as sea levels rose and fell.

New excavations and analysis of the Boodie Cave site on Barrow Island—which was part of the Australian mainland at the time the site was occupied—shows that humans had reached that point by perhaps 51,000 years ago. The archaeological finds include stone and shell artifacts, charcoal, and the remains of shellfish, sea urchins, fish, turtles, and other vertebrate and invertebrate fauna. An investigative team led by Michael Bird refers to this lifestyle as a *maritime desert adaptation:* one that successfully exploited both coastal and adjacent desert habitats.[9]

The Bird team's analysis of sea floor geography and vantage points demonstrates two key facts. First, contrary to previous thought, some islands in the Sahul Banks chain were visible from some places on Timor and Rote Island at about 55,000 years ago. The people who set off to see what that new place was like knew there *was* another place. If they could not actually see the next island, they could probably see cloud formations that revealed its presence. Getting to Australia was likely to have been an exploration, not a catastrophic accident.

Second, the team calculated that the probability of canoes or rafts drifting accidentally from Timor or Rote Island and landing on the Australian mainland was very small, based on simulated drifting voyages. On the face of it, the chance that enough people to form a long-lasting population were swept to Australia in close geographical and chronological proximity without having good maritime skills seems small. If the colonization of Greater Australia was accidental, what must be postulated ranges from the "pregnant woman on a log" hypothesis to the idea that small fishing crews or other seafarers on crude vessels, such as dugout canoes or rafts, were swept to Australia by storms, tsunamis, and other relatively uncommon events. Therefore, the Bird team suggests that migrations between the islands of Island Southeast Asia were deliberate and facilitated by paddling.

Interisland trips are very common among seafaring peoples. Though the required open sea voyages may have prevented archaic humans from getting to Greater Australia, once modern humans mastered navigation, seamanship, boat-making, fishing, and the making of twine, nets, or rope, such voyages would have been part of everyday life. The ability to plan and execute such a journey to an unknown territory clearly involved deep foresight and the ability to conceive of complex, multipart technologies. Further, building a useful form of watercraft would involve using technological components derived from different places at different times and solving or withstanding difficulties that had never been experienced before.[10]

This interpretation of the peopling of Australia is partially supported by the evidence of pelagic fishing, extensive marine exploitation, and use of shellfish and marine resources that was found at the Asitau Kuru and Lene Hara sites on Timor-Leste. Although both are dated to about 42,000–35,000 years ago—more recent than the occupation of Boodie Cave on Barrow Island—they demonstrate that a maritime adaptation in this location required no artifacts that were very different from those in older sites. Once the requisite technical skills, planning abilities, and cognitive skills had been acquired, water was not a barrier but an invitation to explore different coastlines, different reefs, or different deep ocean areas.

NINE

Adaptations

Once modern humans had the maritime skills to reach Greater Australia, I maintain that the acquisition and retention of detailed knowledge of the geography and natural resources in Greater Australia—rather than any particular technological advance—was the key to human survival. They had no dogs.

The most crucial knowledge the First Australians would have had to master in order to survive in Sahul was the location of both ephemeral and permanent water sources. A survey by Michael Bird and colleagues has shown that all fifty-five known archaeological sites in Australia older than 30,000 years are within forty kilometers, or two days' walk, of a permanent water source. Fully 84 percent of the sites are within a single day's walk of water. Knowing where the nearest water is might make the difference between dying and surviving, and these remarkable numbers argue for a deliberate placement of living sites based on geographic knowledge and a keen appreciation of how long humans can go without water. In a larger sample of 1,049 more recent sites—those less than 30,000 years old—only 65 percent are within a day's walk of water. This implies detailed, very specific knowledge. The team argues: "Aridity in isolation therefore is not necessarily a barrier either to habitation or to transit. It is the duration of inundation, the connectedness of water at times of inundation, and the location of permanent water in the landscape that dictates where, and for what length of time, humans could reside in or transit through most of interior Australia."[1]

Another way to look at the settling of Australia is to ask: How long after initial landfall did the First Australians learn enough to be able to inhabit the arid central interior? This is probably the most difficult environmental zone for humans to inhabit, even today, because it gets less than 200 millimeters of rain annually and permanent water sources are scarce. Warratyi, a rock shelter that preserves megafaunal remains and tools dated to around 49,000 years ago, is the oldest securely dated archaeological site in the arid central zone of Australia. If the dates are correct, there was a long 16,000 years between the oldest layers at Madjedbebe and the oldest levels at Warratyi. The uppermost levels at Madjedbebe that bear artifacts are only 18,000 years old, so the period of occupation at Madjedbebe nearly spans the time at which humans were able to live in the arid zone. The Warratyi rock shelter preserves evidence not only of humans but also of cultural innovations, such as very early ochre use, gypsum use, bone tools, hafted tools, small artifacts that were blunted along one edge and often hafted to some sort of handle or larger object, and emu eggshell. The shelter also preserves evidence of the hunting of Australian megafauna such as *Diprotodon*, a wombat relative that was about the size of a hippo and weighed roughly 2,790 kilograms. The finds were in a stratified site dated by optically stimulated luminescence (OSL), dating of quartz grains, and radiocarbon (carbon-14) dating of hearth charcoal and avian eggshells (probably from the giant flightless bird *Genyornis newtoni*).[2]

Another site suggesting early arrival of humans in Greater Australia is the Nauwalabila rock shelter in Arnhem Land, fairly close to Madjedbebe. A combination of carbon-14 and OSL dating conducted at Nauwalabila yielded a probable occupation date of more than 50,000 years ago, most probably about 57,000 years ago. James O'Connell and James Allen questioned the dating on the grounds that the artifacts may have migrated downward—to lower, older sedimentary levels—in the site, thus indicating a falsely ancient date. There is no evidence to refute or confirm this possibility. O'Connell and Allen have often been skeptical of early dates on Australian sites.[3]

Skeletal remains from another famous site, Lake Mungo, were initially thought to be even older, dating to 62,000 years ago, but later studies using OSL dating revised this assessment. The Mungo 3 burial of a skeleton is now generally conceded to be about 42,000–40,000 years before present. Though this is called the Willandra Lakes Region, and the most important site is known as Lake Mungo, there are no permanent bodies of water there now. There are sand dunes thought to have been deposited by a now dried-up lake. Other Pleistocene sites from the area contain shellfish, fish, and crayfish, along with a smaller component of small to medium-sized mammals. The restricted size of the fish bones suggests the use of nets. Initial claims of great antiquity for the Mungo 3 skeleton and the Mungo 1 cremation of a second individual from New South Wales suggested very rapid movement within Australia from northwest to southeast, but those dates have been revised. In any case, even if the First Australians had traveled through the central desert—an extremely difficult, even dangerous terrain for humans today—this route would have necessitated a migration or territorial expansion of more than 2,000 kilometers after landfall, including adapting to different habitats and resources along the way. Following a less forbidding coastal route would have involved a much greater distance, but fewer new circumstances along the way.[4]

Another key site, Devil's Lair in southeastern Australia, is also a very long way from where landfall occurred. Dated to 48,000 years ago, Devil's Lair offers a continuous Pleistocene sequence containing animal bone, artifacts from a small flake-based industry, and charcoal. The abundant fragmented and charred bones indicate a heavy reliance on terrestrial hunting of southern bandicoot, bettong, and kangaroo. The former two species could have been netted or caught in traps, but the kangaroo is a larger animal that almost certainly required some sort of distance weapon.[5]

The edge of the western Australian arid zone has yielded several sites dated to 40,000–35,000 years ago (e.g., Mandu Mandu, Jansz, C99, Pilgonaman), showing evidence of coastal habitation by remains of marine mollusks, crabs, and sea urchins as well as small land mammals such as wallabies. However, Warratyi rock shelter, at 49,000 years old, is currently the oldest stratified site recording human occupation of the arid central zone. The artifacts found there demonstrate a thorough adaptation to the environment.[6]

As telling as these early sites is the evidence of the earliest known ground-edge stone axes. For a long time, Carpenter's Gap 1, a site in northwestern Australia, had the oldest known ground stone axes made from basalt, which are at least 49,000–40,000 years old. The analytical team led by Peter Hiscock forestalled potential criticisms by a careful series of ex-periments and studies. To prevent mistaking artifacts in situ from those that might have migrated downward in sediments, the team looked for a relationship between the size or mass of the axe fragments and their position in the stratigraphic se-quence. None was found. Analysis of the manufacturing traces on the axe fragments indicated that abrasion techniques were extensively used and the artifacts were ground and smoothed after shaping. Such axes are unknown in roughly contempora-neous archaeological sites on islands to the northeast of Greater Australia, indicating that ground-edge axes were invented at or shortly after landfall. Technological innovation may have been triggered by needs posed by new environments.[7]

Other sites with this new technology occur at several northern Australian sites, including Malanangarr, Widgingarri 1, Nau-wabila, Nawamoyn, Nawarla Gabarnmang, Sandy Creek, and now—much earlier than any of the other sites—Madjedbebe. Very recently reported excavations on the island of Obi, on the Maluku archipelago of Wallacea just west of the Bird's Head Peninsula of New Guinea, have uncovered beautiful ground stone axes. Although the sites were first occupied about 17,500 years ago, the earliest evidence for axe making begins about 14,000 years ago, and the earliest specimens are made from large clam shells, not stone. Flakes from stone axes appear about 12,000 years ago and may signal more intensive forest clearance as the climate warmed and forests closed in. As this is the first excavation on Obi, and a small excavation, it remains to be seen if there are earlier sites with ground-edge stone or shell axes or adzes.

The development of such axes and adzes marks the inven-tion of a new technology that is exceptionally difficult and time-consuming to produce—a technology that may have been instrumental in crossing Wallacea. The island is located on one of the least-cost routes for migration from Sunda to Sahul and is densely forested. Ground stone and shell axes are known ethnographically from this region and were used in clearing forested areas, heavy-duty woodworking, or making water-craft. The primary terrestrial animal whose bones are found at the Obi sites is the cuscus (*Phalanger rothschildi*), which may have been indigenous to the island.[8]

Further, evidence of waisted stone axes has been recovered from Bobongara and several sites on the Huon Peninsula and in the Ivane Valley of Papua New Guinea at 50,000–40,000 years ago or earlier. These axes appear to have been used in wood-working and clearing of land, judging from the wear marks from use. Here too, new tools may have been created to cope with a new environment.

Carpenter's Gap 1, in addition to preserving stone axes, has also yielded an ochre-covered slab and a pointed bone imple-ment that is more than 46,000 years old, making it the oldest shaped and utilized bone implement in Australia. Based on eth-nographic collections and microscopic use-wear marks, this unusual implement is most probably either a nose ornament or a bone awl.[9]

Putting information together from various sites, Peter His-cock, Sue O'Connor, Jane Balme, and Tim Maloney observe, "Geographic variation and regional traditions of behaviour are evident in the technology of humans colonising Sahul. This is epitomized by the Pleistocene use of hafted ground-edge axes in northern Australia, and flaked/waisted unground axes in Papua New Guinea, but no axes at all in the southern two-thirds of Australia." In other words, new tool technologies and regional variation in tools are present from almost the very beginning of human habitation in Australia—including at the oldest known site—although the likelihood is that all of the early migrants into Greater Australia came from a single population. As people spread from landfall into new habitats, they modified their technology to suit local needs.[10]

For the same reason, linguistic variability was also likely to have been limited at landfall, although analyses of the languages on Australia and New Guinea at European contact reveal tremendous variability, including more than 1,200 languages. Both linguistic and technological lines of evidence suggest that humans reached Greater Australia only once or a very few times. However large the initial population was, the people apparently split up into smaller groups almost immediately, perhaps to improve their chances of survival in an unknown place. The number of early archaeological sites in quite different habitats and in quite different parts of Greater Australia probably reflects a series of small, transient, and mobile groups. Landfall in Greater Australia at 65,000 years ago (or, more conservatively, some time before 50,000 years ago) is thus a reasonably well-supported conclusion. Of course, no one realistically expects to find an archaeological site that was formed by the very first incomers to set foot on the Australian continent, but the evidence from Madjedbebe and the chronological cluster of sites from about 55,000–45,000 years ago seem congruent. Too, this reading of the evidence supports genetic evidence that suggests an early split occurred between modern human populations that migrated to Eurasia and those that spread to Greater Australia, and that a later division occurred between Australians and Papuans at between 70,000 and 51,000 years ago, with a best probability of about 58,000 years ago.[11]

Landfall for humans on the northern coast of Greater Australia seems geographically more feasible than landfall elsewhere because the open-sea voyage at the start would have been shorter. This idea agrees with the location of the oldest archaeological site on the mainland, which is currently Madjedbebe (previously called Malakunjala II). This is a multilayered, stratified site, with stone tools and hearths from the top to the lowest level. The lowest archaeological layers are too old and the organic materials too degraded for accurate carbon-14 dating, but OSL indicates that the beginning of the occupation of the site dates to between 70,700 and 59,300 years ago, with the highest probabilities centering on 65,000 years ago. Madjedbebe is not only the oldest currently known archaeological site in Australia, it has also yielded evidence of the oldest ground stone axes in the world.[12]

To be fair, serious questions about the antiquity of the lowest levels of Madjedbebe have been made by a number of scholars, in part because there are no other sites of such known antiquity in Australia. That, of course, is a flimsy argument; whether its identity is agreed on or not, some site has to be older than all others. But I will review some of the objections here so readers can consider their validity. The objections center on the idea that artifacts, charcoal, or other objects that might bear on the age of the occupied site may have moved downward to more ancient levels within the sediments. Carbon-14 dates are reliable up to about 50,000 years ago; older dates rely on OSL dating of sediments to determine when they were last exposed to sunlight. If artifacts are no longer associated with the sediments that originally buried them, the resultant dates will be wrong.[13]

James O'Connell is forceful in his questioning of the Madjedbebe dates. The primary suggested cause of artifact movement is bioturbation, the disturbance of soils caused by the actions of biological agents, particularly termites. Termites infest many tropical soils, and three species of termite probably lived in the region of Australia where Madjedbebe is found. Termites burrow through soils to acquire sand grains and a fine slurry of clay mixed with saliva, out of which they build the termite mound on the surface. The repeated removal of fine sand grains concentrates coarser particles into a gravelly sub-surface layer that is sometimes called a stone line or stone layer. As they mine the sediments for fine particles, termites are also searching for plant material to eat. Termite mounds may be as tall as seventy centimeters and in general last about three years before they break down and are abandoned. Abandoned mounds are eroded by intense rainstorms that move fine soil particles downhill. Disintegrating termite mounds yield soil particles that may move downhill very slowly because of slope wash and gravity. When termite tunnels are no longer in active use, they collapse and become invisible and undetectable.

Thus far, there is no positive evidence of the ancient presence of termite tunnels or mounds that has figured in the criticisms of the dating of Madjedbebe. No ancient termite remains, fecal pellets such as usually line termite tunnels, or remains of termite mounds have yet been recovered from there. Madjedbebe is indeed in the broad geographic region in which termites live or lived, but that fact does not provide evidence that there was termite activity in the precise region of the archaeological site in the past. The crux of the disagreement between those groups favoring the activity of termites and those asserting there has been little disturbance of the site is this: the two sides cannot agree on which criteria can be used to recognize termite bioturbation and on how severe that bioturbation's effect on the dating would be. To argue convincingly for termite activity, the skeptics must produce evidence not only that termites could have produced some of the effects observed, but that they did produce them. Martin Williams, in a series of recent papers in collaboration with O'Connell and others on possible termite activity, wrote: "We do note, however, that termite presence will not always result in significant artifact displacement and stone layer formation—these outcomes require prolonged high-density termite populations, such as found in the tropical north of Australia. . . . In lower termite density regions the outcomes will be more situation-specific."[14] Those arguing against termite bioturbation should continue to pursue sound criteria for detecting the presence or absence of termite activity.

All phases of occupation at Madjedbebe include numerous remains of charred macrofossil plant remains, not including wood but including more than 1,000 specimens from a hearth feature at the lowest level. Williams and his colleagues have suggested that these features are incorrectly identified as hearths, but charred plant food remains would seem to be fairly definitive evidence that they are cooking hearths. Remarkably, some of the plants from the site required extensive processing to be edible, a display of detailed knowledge some 23,000 years earlier than at any other site. Skeptics are disturbed by the presence of plant foods requiring substantial processing so early in time. For example, some remains are those of palms (Arecaceae). Though the apex of the pith can be eaten raw or lightly roasted, most of the pith must be roasted for about twelve hours and then laboriously pounded to remove fibrous elements, thus making the starchy carbohydrates accessible for consumption. This interpretation of plant processing is supported by seed-grinding stones in the artifact assemblage. Other plant foods found include tubers, fruits, nuts, and seeds that demonstrate a broad diet and great adaptability in exploiting new foods. The presence of charred food remains requiring considerable preparation also attests to the preservation of very fragile and delicate specimens that might not survive extensive bioturbation.

With the recent work at Madjedbebe and other sites in Australia indicating human landfall as old as or older than about 50,000 years, evidence for the peopling of Australia earlier than 50,000 years ago is mounting. There is also firm evidence that the First Australians did not stay near landfall, as might be expected, but spread out to all corners of the continent and to a highly diverse set of habitats. Within a few thousand years of landfall, and possibly sooner, the First Australians left archaeological sites in the savannah, the tropical rainforest, open woodlands, in the zones between woodland and scrub and savannah and the arid zone, in coastal habitats, and in the steppe environments of Tasmania, the highland valleys of New Guinea, and the Willandra Lakes Region of New South Wales.

More work on dating accuracy and possible causes of chronological distortion is certainly warranted. What remains remarkable is that newly studied sites pushing the initial landfall toward 50,000 years ago or older also keep yielding evidence of early innovations, new tool types, uses of new resources, and sophisticated adaptations to new environments. Greater Australia is full of surprises.

TEN

Surviving in New Ecosystems

WHAT HAPPENED during the settling of Eurasia was different from what happened in Australia. Different challenges made life difficult; different local factors led to success in each.

The future inhabitants of Eurasia simply pushed north and east into new areas and new territories. Perhaps they were following game and adjusting gradually over time to differences in climate and fauna without facing any startlingly new challenges or long individual journeys. The biggest changes that would have affected their hunting success were the colder weather and the association of human hunters with canids, which allowed them to take down more of the megafauna using new and more efficient hunting techniques. Dogs both improved hunting success and made it easier to detect other carnivores trying to steal carcasses from the hunters.

Who exactly those other carnivores were is important. The Ice Age African and Eurasian predatory guilds had many felids:

the lion had an average male weight of 190 kilograms (kg) and an average female weight of about 127 kg. The now-extinct cave lion was even bigger, with males at 318–363 kg. There were also tigers, leopards, large saber-toothed cats, and cheetahs, plus another ten species of medium-sized Eurasian cats (including two species of lynx), the clouded or snow leopard, and seventeen species of smaller wildcats. Africa and Eurasia also had three species of wolf as well as the African wild dog, three species of jackal, the dhole, and numerous smaller species classified as various kinds of fox. There were also hyenas (three species at least), honey badgers (something like wolverines), and various small mongoose species including polecats, otters, civets, and weasels. Together, this varied array of carnivores preyed on herbivorous species ranging from the very small—mouse-sized—to the very large, in the form of elephants, rhinos, hippos, African buffalos, and giraffes. No carnivore was as large as the adults of the largest prey species; the larger species often forced a group of predators to work together to bring them down. Juveniles of a prey species could also be hunted; they were physically smaller and had less stamina than adults, and might also be less wary of potential predators. However, prey animals often clustered together for mutual protection.

The Australian fauna was strikingly different. There were no felids or other catlike mammals at all until Europeans arrived in 1788, bringing house cats with them. Feral cats in Australia are now large compared with most house cats, and they are also widespread. At the time of European colonization, Australia had only one doglike predator, *Thylacinus cynocephalus,* a marsupial. Thylacines hunted both singly and in packs and were notable for their stamina in chasing down prey. They were much smaller than the biggest Eurasian predators and

This photo of a living thylacine in the Gardens of the London Zoological Society was taken in the 1910s. The stiff, striped tail, striped rump, and huge gape, combined with its generally doglike appearance (though it was a marsupial), were of great interest to visitors.

weighed about 14.5–21 kg, with males being larger than females. Those now-extinct thylacines—known colloquially as Tasmanian tigers, striped hyenas, or Australian wolves—might have competed directly with humans. The thylacine was extinct in Australia as of September 7, 1936, when the last, poorly cared-for specimen in Hobart Zoo in Tasmania died. There had once been several other species of thylacine in Australia, dating back to the Miocene, that appeared to be better adapted to forest conditions than was the modern, generalized thylacine. *Thylacinus cynocephalus* had to contend with an overall drying trend and with other ecological changes, including the arrival of people and dingoes, and went extinct in less than 150 years after Europeans arrived.[1]

The oldest directly dated dingo specimen is a burial on Timor-Leste that is about 3,000 years old—much more recent than the First Australians. A dingo weighs 10–15 kg. Male dingoes are larger than females and also larger than female thylacines, which they may have preyed upon. Dingoes never reached Tasmania, which probably contributed to the fact that thylacines survived later there than on the mainland. Despite reports of sightings of thylacines from time to time, expeditions to find surviving thylacines in Tasmania have been as yet completely, and sadly, unsuccessful.[2]

Very little is known about the thylacine diet in the wild. As Robert Paddle and David Owen have pointed out in separate books, the natural feeding habits of thylacines have been much written about but are very poorly documented. Though thylacines were known informally as sheep-killers, there are astonishingly few instances of eyewitness reports of such activity and many exaggerations. Historical documents far more often record sheep killing by free-ranging domestic dogs than by thylacines. In Tasmania, substantial bounties for live thylacines, or for their skins, in the 1800s brought in very few individuals. A similar lack of documentation undercuts the once-popular rumor that thylacines were blood-suckers that killed sheep, sucked out their blood, and left the carcasses otherwise untouched. That story smacks more of European vampire mythology than reality.

The extinction of thylacines on the Australian mainland may have been part of the megafaunal extinction that included the end of the biggest indigenous Australian predator, the so-called marsupial lion (*Thylacoleo carnifex*). Many marsupial species went extinct about 40,000 years ago.[3] Sadly, no one in living memory has ever seen a live specimen of *Thylacoleo carnifex*. Although the First Australians must have seen them, they left little record of the marsupial lions' habits or prey. There are two rock paintings identified controversially as depicting *Thylacoleo*, but this interpretation is not widely accepted.[4] In body size the marsupial lion ranged up to about 150 kg, roughly the size of a leopard or African lion. It is very difficult to identify its predatory habits and carcass destruction patterns with confidence because we have no samples of bones known to have been damaged by this particular extinct species. However, the carnassial teeth of *Thylacoleo*—the ones used for slicing up meat—are downright extraordinary in size, the largest relative to body size of any mammal.[5]

Thylacoleo's extraordinarily specialized slicing teeth suggest it was a hypercarnivore, a species whose diet consists of more than 70 percent meat. As is common among hypercarnivores, *Thylacoleo carnifex* had greatly elongated blades on its slicing cheek teeth, the carnassials. It was able to generate exceptional bite forces with those teeth, which were probably used in a rapid killing or windpipe-crushing bite.

Thylacoleo was adapted to be a robust ambush predator that lay in wait for prey, maybe in a tree from which it jumped to attack. Its build was not that of an animal that chases its prey as wolves do. It had powerful forearms and a semi-opposable thumb ending in a large, sharp claw for seizing its prey. The current consensus is that *Thylacoleo* probably preyed on adults and juveniles of all but the very largest herbivores in Greater Australia. On the basis of an analysis of marks on bones from Lancefield Swamp, Horton and Wright proposed that *Thylacoleo* was responsible for all the apparent tooth scratches in that assemblage. The location of the marks is congruent with the idea that marsupial lions were meat eaters,

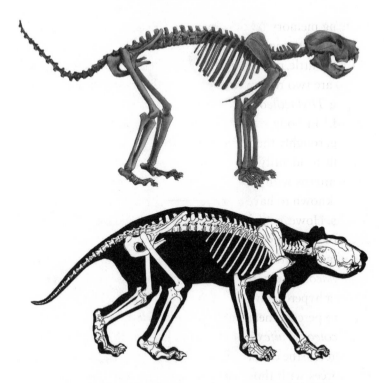

Unfortunately there are no photos of the largest marsupial predator, *Thylacoleo carnifex,* sometimes called the marsupial lion. Working from fossil remains, Rod Wells and Aaron Camens reconstructed a complete *Thylacoleo* skeleton (*top*); Peter Murray created the body silhouette (*bottom*).

not bone crushers. Whether the first humans in Australia deliberately hunted *Thylacoleo* as a dangerous competitor is unknown, but marsupial lions disappeared about 40,000 years ago, well after human arrival.[6]

A third carnivorous marsupial in Australia, represented by the Tasmanian devil (*Sarcophilus harrisi*), still survives. It is considerably smaller in body size than thylacines or marsupial lions. From 51–79 centimeters in length, the Tasmanian devil weighs from 4 to 12 kg. This means a Tasmanian devil is smaller than the English bulldog that many modern humans keep as a pet. Though it is not large, the Tasmanian devil is ferocious and very aggressive. It has strong jaws and stout teeth to crush the bones of scavenged carcasses. It also has a number of much smaller relatives, such as the numbats, quolls, and dunnarts, which eat insects, small rodents, fruit, and lizards. The small carnivores are unlikely to have posed much competition to humans.

That's it—at human arrival, there were only two medium-to large-sized mammalian predators in all of Greater Australia. Lest the continent sound too easy to live in, remember that what flourished instead were reptilian predators somewhat akin to Komodo dragons, large monitor lizards or goannas, and very large and sometimes poisonous snakes, crocodiles, and a few enormous, flightless birds. There were not nearly as many prey species or predators as in Africa or Asia.[7]

Why was there a paucity of terrestrial predators and prey in Australia? It has been argued that this is attributable to Australia's relatively small landmass compared with Asia, Africa, or Europe. By landmass alone, Asia is nearly five times larger than Australia, Africa is more than three times larger, and Europe is one-and-a-tenth times larger. What matters as much as sheer landmass is the proportion of habitable land and its productivity. The arid climate of central Australia means that much of its land is difficult, if not impossible, to inhabit for long.

Though life was sometimes precarious for the First Australians, there is no denying their huge success in moving into this new place. Predators that hunted the same medium-sized animals that humans did were fairly rare, so competition was not extreme compared with what the humans who moved into Eurasia faced. What was extreme was the sheer novelty of Australia compared with Africa and Eurasia. Australian species that demanded to be taken seriously as predators included not only the three marsupial carnivores but also large predatory birds and immense reptiles, not to mention plants, spiders, snakes, toads, and jellyfish that could prove fatal.

What effect did the unusual indigenous fauna of Australia have on human success in the new continent? I argue that the peopling of Australia is one of the most important and telling migrations or territorial expansions carried out by modern humans. It posed entirely new challenges for the people who became the First Australians. Australia was at best a difficult place to survive that needed special skills and adaptations for survival. What the archaeological and ethnographic records tell us is that what was needed more than better weapons or specialized tools was knowledge.

What the Australian story tells us is this: humans can and will adapt to all kinds of new and difficult conditions by acquiring and sharing knowledge. Knowledge can compensate for a lack of material goods and may spur the creation of new ones. What the First Australians brought with them to Greater Australia was seemingly scant in terms of the material culture that has survived, but the ability to learn, to notice, to remember, and to share information was of enormous importance. This is nicely summed up by an aphorism from the Aranda-speaking people of central and northern Australia, often quoted by my archaeologist friend John Shea: "The more you know, the less you need."

ELEVEN

Why Has the Australian Story Been Overlooked for So Long?

IF THE AUSTRALIAN settlement record matters, and I believe it does, why has it been overlooked for so long? The peopling of any continent is of great interest to those who live or work there, but perhaps of somewhat less interest to people elsewhere. Nonetheless, understanding migrations and adaptations has huge potential to enlighten us about the nature of the human species itself and the role of humans in various ecosystems. The peopling of Australia is a story that has been too often neglected by the rest of the world as uninteresting or uninformative, neither of which is true.[1]

The lingering taint of prejudice can also be seen in another frequently used technique to track and understand human migrations: genomic studies. Deciphering the genetic code of humans or other animals usually relies on ancient DNA (aDNA) extracted from preserved bones and teeth to establish relationships. Scientists also study modern DNA to obtain and compare genetic information among species. If you believe that Indigenous Australians are isolated and boring, why look at their DNA? Of course, this means ignoring potentially important information and tacitly endorsing the harmful racial prejudices of the eighteenth, nineteenth, twentieth, and twenty-first centuries.

One standard approach in genomic studies is to sample the DNA of two extant groups and infer the time of their separation from a common ancestor by the number of differing mutations between the two samples. Another is to sample aDNA and compare the samples with modern genomes to try to identify the now-living group to which the ancient person was most closely related. In both cases, a key problem is how many individual samples are needed to fairly represent the diversity that did once or does now exist. Another serious difficulty is that living people often do not know their ancestry in great detail and may self-identify with a particular ethnic or geographic group without knowing whether their ancestry includes people from other groups. Once European colonization began in Australia around 1788, the incidence of genetic admixture—both forcible and voluntary—among different human groups greatly increased but was rarely openly acknowledged. Determining which genetic sequences identify one group or another is very difficult, and the answers sometimes contradict family history in disturbing ways.

Many of the Indigenous Australian women who worked on sheep or cattle stations or on missions formed sexual relationships with men of European heritage and had children with them. The nature of these situations varied widely, and the parentage of many children went unacknowledged. Between the 1910s and the 1970s, it was official government policy to remove Aboriginal and Torres Strait Islander People's children from their families and to place them in institutions—often mission schools—to be educated, in the hope that they could learn Western ways and pass for people of European heritage. These children became known as the Stolen Generation. It is estimated that as many as one out of every three Indigenous children was removed from their family. The avowed aim was to expunge their Indigenous qualities, force them to assimilate, and then wipe out their culture. Many of these children were cruelly forbidden to speak their native languages or have contact with their families. Many were sexually abused, beaten, or simply traumatized by being removed from their families and culture.

Even if DNA is not admixed, it may be degraded to the point of uselessness by natural forces. Once DNA is extracted and decontaminated, interpreting the genomic information correctly can be problematic. Until very recently, most "worldwide samples" or data banks of "global genomes" of humans and other species contained few or no samples from Greater Australia, modern or ancient. In part, this deficit can be attributed to a dismissive assumption that the Australian story will simply not be informative. Historically, starting with the era of colonization, Indigenous Australians were regarded as an unchanging, uninventive people with a primitive culture who were almost frozen in time. Laypeople of European ancestry were basically the only ones who collected stone tools or other artifacts for museums or private collections, who inspected encampments, or who showed any interest in Australian prehistory until the field was revolutionized beginning in the 1960s. It was not until this late date that systematic, stratigraphic excavations of Australian sites began to be conducted professionally, establishing a time, depth, and spatial distribution of sites and artifacts rather than simply collecting curiosities on or near the surface.

This dismissal can also be partly attributed to unsuccessful attempts to obtain DNA samples from Indigenous peoples who have historical reason to be mistrustful of the actions of Europeans. Robert Hughes, in his masterful book *The Fatal Shore*, records that the first words known to have been spoken to white Englishmen by Indigenous Australians were "Warra warra!" The words meant "Go away!" and were shouted by a group of men who were shaking spears at the foreigners upon their first sight of the English ships entering Botany Bay. If this story is accurate, it would seem that the First Australians felt threatened by the invading European strangers from the beginning. In many senses, they were right in assuming that outsiders would do them harm.[2]

Many Indigenous peoples today are wary of exploitation by outsiders. Religious or spiritual beliefs may make them reluctant to condone the removal, damage, or display of the bones of their ancestors, or of sacred objects, or of sacred places; they may also object to the analysis of bodily tissues or fluids. Establishing trust and communication between scientists and people who either donate or permit the use of places or items regarded as powerful is vital. Because Indigenous beliefs and knowledge may include critical information for interpreting the past, scientists and Indigenous peoples may find that collaborating is highly beneficial to both.[3] If we are to understand variability in human or animal genomes, lifestyles, or adaptations to different ecosystems, we cannot leave out any group of humans, and certainly we cannot leave out a group whose archaeological record promises to reveal more about the abilities and adaptations of early modern humans than any other. However, it would be equally foolhardy to assume that today's Aboriginal peoples live as their ancestors did thousands of years ago. Their anatomy and genetics may remain largely the same as they were in the past, but which abilities, behaviors, beliefs, or knowledge Indigenous Australians exhibit today does not tell us much about what they did in the deep past. Only the archaeological record is a reliable source for that, and it provides a sketchy outline at best.

The professional prehistorians and archaeologists who first looked at the cultures of Australia were heavily conditioned by their elders in Europe to look for a sequence of tool types that could be used as a means of dating deposits. If, for example, there were Acheulian tools at a site, then the site had to be older (prehuman) than sites with Mousterian tools—typically associated with Neanderthals. Mousterian tools, in turn, preceded the Levallois tools shaped by more modern humans. This is no longer seen as unvaryingly true in Europe, and it certainly does not apply to sites in Greater Australia, but such a story was the expectation in the late eighteenth, nineteenth, and early twentieth centuries. Robert Pullein, in the early part of the twentieth century, pronounced that Indigenous Australians were "an unchanging people in an unchanging environment." This damning sentiment was basically echoed by the internationally revered prehistorian Vere Gordon Childe, who regarded Australian archaeology as "horribly boring" because there seemed to be no innovation or progress, and he declared Australia itself to be a "cultural backwater." He and his contemporaries did not see, as modern archaeologists do, that Indigenous cultures adapted both cleverly and inventively as the climate changed and they undertook migrations to different ecosystems; they did not remain static. Iain Davidson, representing the more modern approach to archaeology, has aptly pointed out that "the history of people in Australia shows how they were *a changing people* in *changing environments,* adapting in remarkable ways to the challenges and opportunities of their surroundings."[4]

The Australian record is important in its unusual circumstances. Because of the geographical isolation of Greater Australia and the harshness of much of its geography and climate, the evolution of species once they reached the continent has not been heavily influenced by the arrival of other species from elsewhere or by recurrent migrations. The record of the past and prehistory of Greater Australia seems more straightforward and less confusing than that of Europe or Asia, though of course our knowledge is far from complete. This remarkable story of a continent almost completely isolated from animals or plants from elsewhere is unparalleled. Even genetic material from living Australians shows few signs of admixture with Asians, Pacific Islanders, or Native Americans—but there are some. Modern humans and the native fauna and flora have lived in Australia for at least 60,000 years with very few intrusions from outside until perhaps 5,000 years ago, when dingoes appeared.

Like humans, dingoes or their ancestors must have been transported by boat, but we don't know why or by whom, as the people of the boats left few informative traces of their origins. But to trace the evolution and impact of domestication on dogs and their human companions, we have to admit that dingoes may help to illuminate the state of canids at or prior to domestication, and so we must link the 60,000- or 65,000-year-old human arrival with the 4,000- or 5,000-year-old dingo arrival to make a coherent whole.[5]

The story of Australia is especially significant because it provides a valuable instance of the events involved in the initial peopling of a continent by modern humans and the consequences of that peopling. One of the primary handicaps in trying to decipher what happened when modern humans moved out of Africa into new geographic areas is that, if we focus on only the Eurasian story, we are dealing with a data set consisting of only a single sample. This is frustrating and may lead to false conclusions.

The peopling of Australia is neither a minor side issue in the record of human migration and adaptation, nor a not-very-important question to be resolved when the bigger questions have been answered. The peopling of Australia is not so much a question as part of an answer to broader inquiries about human migration, adaptation, and evolution. What happens when people move into new areas, new ecosystems, new climates? How do they adapt? Why do humans tend to bring their dogs with them when they expand their territory? How did it affect the First Australians not to have dogs?

New evidence about the earliest archaeological site in Australia has in fact shaken up the global narrative. Modern humans' entry into Australia was the first instance of their expansion into a huge new land mass. It may have occurred as early as about 70,000 years ago or as late as about 55,000 years ago; if so, it would have preceded the modern human expansion into central and eastern Europe at about 50,000 years ago. The split between future Europeans and Asians, and future Australians, must have occurred earlier still. If we want to understand how such splitting of a population occurs, we ought to be taking the Australian story as a model and first instance. The record of humans entering Europe and Asia, long given primacy in the scholarly and popular literature, is in fact the second instance, a narrative of human migration and adaptation that offers another glimpse of how such events occurred and varied under different conditions.

An important aspect of the initial adaptation to Australia was the sharing of knowledge. The teaching and the learning of the knowledge of any local region or "country" is deeply embedded in the traditional cultural practices of Indigenous Australians. As the First Australians moved south and east, possibly taking a coastal route to Greater Australia, they developed some novel and impressive mechanisms for exploiting coastal resources. (As far as we know, none of their technological advances involved working with domesticated dogs.) Through the millennia, climate changed; ocean levels rose and fell and shorelines altered, meaning that parts of what had once been Greater Australia became islands, such as New Guinea and Tasmania. There was still plenty of land and plenty of game, but none of it was familiar.

TWELVE

The Importance of Dingoes

BECAUSE CANIDS were the first animals to be domesticated, and animal domestication transformed human life, tracing the location, timing, and function of canid domestication is important. As far as we know, dingoes were only the second large placental mammal (after humans) to reach Australia. Like humans, dingoes came to Australia by boat, a fact that implies to some analysts that dingoes were at least semidomesticated when they reached Greater Australia. However, many undomesticated animals have also been transported by boat, such as foxes, giraffes, moose, various cervids, lions, and cheetahs, to name a few. Traditional knowledge, expressed in dances (corroborees) and myths, also asserts that dingoes were transported to Australia—accidentally or purposefully—by coastal boat-using peoples, even if dingoes were not fully domesticated at the time.

Semidomestication is an ill-defined term. True domestication involves human influence over the breeding and survival of the young; people thus exert a selective force over the genetics of the species. Semidomesticated dogs are often called village dogs, which are both free ranging and free breeding. So-called village dogs live primarily by scavenging in human-managed environments. They are more tolerant of humans than truly wild or feral dogs and may have socially recognized owners and names. Though village dogs depend on human food and scavenged remains, and could be considered commensals of humans, humans do not control village dogs' breeding choices. Such dogs are often admixed with various recognized breeds—even in Asia these are often European breeds—and may be native to a particular geographic area, forming groups similar to landraces of plants. But there is little evidence of selective breeding in Asian village dogs. Genomic studies of village dogs document considerable genetic diversity, which has led some researchers to suggest that they are close to the original type of canid that was domesticated. If semidomesticated village dogs were the direct ancestors of dingoes, then dingoes cannot reasonably be understood as a fully domestic species gone feral after arrival in Sahul. It is tempting to explain the origin of dingoes by invoking the long-standing practice of people coming by boat from Makassar in Sulawesi to the Kimberly (northwestern Australia), and later to Arnhem Land, to collect trepangs (sea cucumbers), an idea favored by Melanie Fillios and Paul Taçon. Traditionally, a group of men would sail to the northwest coast of Australia, set up a temporary village for a few weeks or months, and collect, then dry and smoke, trepangs to take home and sell to Chinese traders as a prized ingredient in food or medicine.[1]

Trepang collectors are known to have started coming to Australia in the 1500s, but they could have come far earlier, bringing village dogs with them. Peter Savolainen and his team documented genetic resemblances between sections of the mitochondrial DNA (mtDNA) of dingoes and of Southeast Asian village dogs. They concluded that the Australian dingo probably originated from a few East Asian domestic or village dogs because they detected only a single mtDNA haplotype in the dingoes in this study, and it matched a dog haplotype that is common throughout East Asia. However, later studies of whole dingo genomes by Kiley Cairns and Alan Wilton showed that the sample used by Savolainen's team was misleading, probably because it consisted of only a few hundred base pairs of mtDNA. Using whole genomes, Cairns and Wilton found twenty haplotypes among their dingoes, undercutting the conclusion by Savolainen's team and making dingo origins less clear.[2] There is no archaeological confirmation of the Makassan origin of dingoes, as no one has yet found archaeological sites in Australia of great antiquity containing distinctively Makassan artifacts or any sign of distinctively Makassan genes among Indigenous Australians.

Why would undomesticated or only semidomesticated dogs have been brought along on a voyage to collect trepangs? Canids might have been brought on these voyages as protection, as potential food, as a source of warmth, as animals who would clean up refuse and waste, or as companions. Typically, canids fulfill many different roles, and it is difficult to discern which reason is the most relevant. It is likely that, on at least some occasions, dingoes ran off into the bush while their human companions returned home with trepangs or other dried food. The primary evidence for the antiquity of this habit of taking canids along while collecting trepangs lies in tales and dances from the Dreaming—the cultural memory of many Indigenous Australian groups—along with the geographic evidence. But genetic evidence of the distinctiveness of dingoes from dogs, and of native Australians from Southeast Asian islanders, suggests that this was neither a common event nor one that happened repeatedly in antiquity.

Despite difficulties in obtaining samples that are not misleading in one way or another, dingoes and New Guinea singing dogs cannot be omitted from canid genomic analyses without distorting the truth about primitive dogs. Dingoes and New Guinea singing dogs are almost universally agreed to be basal (highly primitive) dogs, possibly representing the state of the earliest domestication among canids. In a genetic study, 112 dingoes and New Guinea singing dogs (plus five male Fraser Island dingoes from another study) were found to share twenty single tandem repeat haplotypes clustered in three haplogroups; most dingoes and all New Guinea singing dogs carried the H60 single nucleotide polymorphism mutation apparently unique to these populations.[3]

A more recent study based on the rediscovery of living populations of New Guinea highland wild dogs in 2014 carried out by Surbakti and colleagues established that the New Guinea highland wild dogs form a distinct genetic lineage with dingoes and New Guinea singing dogs in captivity. These genetic resemblances suggest a close relationship, as do the morphological similarities between dingoes and New Guinea singing dogs. In fact, many canids that are not bred according to established pedigree criteria closely resemble dingoes in appearance, such as pariah dogs, Bali dogs, Indian native dogs, basenjis, and other primitive dogs in Asia. During the research for this book, I have come to think of dingoes loosely as "default dogs" because so many indigenous or primitive dogs resemble dingoes. Yes, dingoes are the primary native dogs of Greater

Australia—an indigenous placental species, if you use a loose definition of "indigenous" for an animal that has demonstrably been in Australia less than 5,000 years.[4]

In a sense, the Europeans who "settled" Australia and Tasmania were a second migration of humans into the Greater Australian continent, many thousands of years after the first. The first European settlers did not know about water sources and other resources; even when shown "bush tucker," or wild food, by Indigenous Australians, Europeans did not understand its importance. Colonials often had little respect for the Indigenous Australians' knowledge of survival, even when they were in dire need of food themselves.

These colonial immigrants created their own hardships by failing to learn from local inhabitants, by failing to move on before they decimated local prey species, and by failing to honor Indigenous ownership of hunting grounds. Europeans did not recognize the intentional, knowledge-based management of the habitats in the landscape by Indigenous Australians. In the first enthusiastic descriptions and reports of the eastern coast of Australia, James Cook and the officers on his initial voyage in the HMS *Endeavour* (1770) wrote that the country was diversified with woods, lawns, and marshes. Cook found the woods free from underwood of every kind and bragged that the trees grew at such a distance from one another that the whole country—or at least a great part of it— might be cultivated without settlers being obliged to cut down a single tree. Like the early generations of European settlers, Cook was simply blind to the deliberate and rather sophisticated ways in which Indigenous peoples managed and owned their territories. The diversified habitats Europeans found upon arrival were maintained deliberately by the Aboriginal peoples. Disastrously, the planned pattern of burning and landscaping via fire, which had been practiced by Aboriginal peoples prior to European arrival, was stopped by the European settlers. Convincing work by Bill Gammage has shown that Indigenous Australians regularly used controlled and calculated patterns of burning vegetation to create a mosaic or patchwork of different plants and habitats in order to influence the distribution of both plant resources and animals to their advantage. What Europeans took for "the biggest estate on Earth" and great natural abundance was, in fact, the product of careful land management developed through deep knowledge of the environment. The Indigenous peoples knew when to burn, what to burn, and how often to burn. It was this deliberate land management that enabled Aboriginal peoples' success in desertic habitats.[5]

The First Australians farmed their land without adopting the fences, permanent settlements, domestic animals, and large food storage facilities that Europeans equated with farming. Neither the European-bred colonials and convicts of 1788 nor some of their descendants today have understood that the degradation of the land and ecosystems in modern times was the direct result of the forced cessation of planned burning carried out by Indigenous Australians. The comprehensive botanical and faunal knowledge and understanding of the influence of fires of particular types upon plants and landscapes were the result of complex and sophisticated observations and practices that contributed hugely to the survival of the First Australians. The colonial invaders did not know how to farm or manage a land like Australia and did not learn readily from Aboriginal people. The Europeans moved in with supplies, seed, stock plants, livestock, and manpower, but they did not see the unsuitability of most of their practices for the land they were inhabiting. Most of the convict labor upon which they depended was not drawn from countrymen and farmers, in any case, so there was little experience in agriculture among the European population.

A well-researched example of adjusting to the unusual demands of living in Greater Australia has to do with the effect of the first domestic dogs that were brought to Tasmania (Van Dieman's Land) by English officers in 1803, which might have paralleled the impact of dingoes in Tasmania had dingoes ever gotten there. (The Bass Strait separated mainland Australia from Tasmania long before dingoes reached mainland Australia, so dingoes never reached Tasmania.[6])

Initially, the settlers' success in hunting kangaroos with imported dogs in the grassy plains near Hobart was great because the kangaroos were not alert to canine predators. Kangaroos were described as abundant, with six or seven kangaroos easily caught in a morning by greyhounds. Owning an English hunting dog was the express privilege of civil and military officers. Aboriginal people began attempting to seize kangaroos killed by Europeans on the Indigenous Australians' hunting grounds as early as 1804. By October 1805, the supplies brought from Europe were depleted and could not be replaced by local equivalents, so the colonists anticipated food shortages. The first chaplain of Tasmania, Robert Knopwood, wrote worriedly in his diary in October 1805 that there were only three weeks of flour rations and five weeks of pork rations left in the colony. The governor decided to issue fresh kangaroo meat as part of the rations due every settler, which vastly diminished the rate of scurvy and kept the colonists from dying. The issue of taking kangaroos with dogs on Indigenous Australians' hunting grounds became a serious conflict as early as 1806 as a direct result of the governor's decision.

There soon developed a "kangaroo economy" in the colony—meat and skins could be obtained by anyone who could steal a dog; they were then sold to the government store or consumed privately. The kangaroo economy enabled not only indentured convicts but also bushrangers (outlaws) to start living outside the legal control of the colony. With a dog, a man could escape servitude and live free. The dogs killed the kangaroos, so bushrangers did not even need guns. Some farmers abandoned agriculture to live as bushrangers, causing more problems with food supplies. Knopwood foresaw some of these issues, yet he more than doubled his annual chaplain's salary by selling kangaroo meat and hide to the government store.

By 1806, dogs were more widespread because of natural population increase, theft, and legal exchange. The resultant overhunting caused a marked decline in the number of kangaroos within a day's walk of Port Phillip or Hobart. As bushrangers and hunters went farther from the settlements to seek more prey, and to escape the regulation, punishment, and control of life within settled colonies, they intruded ever more deeply into Indigenous peoples' hunting grounds; reprisals and murders on both sides became more common. Dogs and the kangaroo economy they enabled triggered the so-called Black War from 1824 to 1832, which was less a war than the brutal and systematic extermination of Tasmanians by colonists.

The geographic isolation of Australia created another useful perspective that might help to trace the migrations and adaptations or failures of modern humans moving into new regions. Geneticist Elizabeth Matisoo-Smith pioneered what is called the commensal approach to investigating the settlement of the Pacific Islands by focusing on the genetic analysis of plants and animals that must have been transported into this area by human agency. In other words, the appearance of new commensal species in a region is a sort of footprint or telltale of human occupation because the new species must have required human transport to get there. Some of the commonly traced commensals or proxies for human presence in Oceania are the Polynesian ti plant (*Cordyline fruticosa*), the paper mulberry (*Broussonetia papyrifera*), taro, and bottle gourd (*Lagenaria siceraria*).[7]

Thus, long before photographs, diaries, and written documents from the colonial settlers recorded the human arrival in Greater Australia there had been another, less formal sort of migration into neighboring Oceania, bringing innovators of the Neolithic revolution. Carried by the Lapita culture, this lifestyle featured garden agriculture, the presence of dogs, pigs, rats, and chickens, and a distinctive style of pottery. This revolution triggered the domestication of local resources and expanded settlement in the Pacific Islands. Pottery in the Lapita style was never widely produced in Australia, nor were chickens or pigs taken up, animals that are important in the traditional culture of New Guineans and various island peoples. And although dingoes are arguably no different from poorly domesticated dogs and thus might usefully serve as proxies for humans, they did not arrive in Australia with the First Australians. Whether dingoes are domesticated Asian dogs gone feral or never-domesticated canids is much debated.[8]

Shao-jie Zhang and colleagues investigated the genetic differences between domesticated dogs and wild canids in the hope of resolving that question. This work relied on the unproven assumption that dingoes were domesticated somewhere in Asia before they were transported to Australia and became feral on the new continent. The fundamental problem with this study is that domestication involves behavioral changes in both the domesticator (humans) and the domesticated species, changes that lead to the preservation of genetic changes if (and only if) humans control the breeding choices of potential domesticates. But there are no signatures of the first behavioral changes in the archaeological record, no tell-tales that reveal when these changes began. Thus, dingoes could be, as Zhang claims, a formerly domesticated species that lost the key behaviors that were acquired during domestication, or dingoes could be a species that was never fully domesticated and was not treated by Indigenous Australians in ways that would lead to domestication.[9]

Zhang's group sequenced ten dingoes and two New Guinea singing dogs, adding these data to ninety-seven canine whole genomes in the literature, including one dingo, forty indigenous dogs from China, Taiwan, and Vietnam, four village dogs from Nigeria, six from India, three from Indonesia, and three from Papua New Guinea, nineteen dogs of various domestic breeds, and twenty-one wolves from across Eurasia. Analysis of these data revealed a clear genetic separation between wolves, two distinct groups of dogs (Indigenous Asian and European), and a group comprising dingoes and New Guinea singing dogs. The researchers found that Indonesian village dogs were most genetically similar to the dingoes and New Guinea singing dogs.

Zhang and colleagues identified fifty genes in dingoes that appeared to have been selected for evolutionarily, many of them related to metabolism of fats and carbohydrates, neuro-development, and reproduction. Thirteen of these differed from the genes found in dogs but resembled those in wolves. Other researchers have found that domestic dogs have a large number of genes that relate to starch digestion, whereas both dingoes and wolves have a very low number of these genes. There is no reliable way, however, to tell when these genetic changes occurred.

Zhang's team estimates that Indonesian village dogs separated from other dogs about 9,100 years ago and that they separated from dingoes about 8,300 years ago, but this jibes poorly with the archaeological and paleontological record. They interpret this finding as showing that dingoes had undergone a genetic reversal during feralization, becoming more similar to wolves, though it is equally possible that dingoes simply retained the initial wolf genes and that the genes were never modified because dingoes were never domesticated.

What about paternal inheritance? Analysis of the Y chromosomes of New Guinea singing dogs by Benjamin Sacks and his colleagues from a highly inbred captive population in the United States suggests that dingoes and New Guinea singing dogs most probably arrived in mainland Australia before the land bridge connecting New Guinea to mainland Australia was submerged. The wild populations of New Guinea highland wild dogs discovered in 2014 are now known to be closely related genetically to New Guinea singing dogs in captivity. Because all New Guinea singing dogs in captivity are descended from eight wild-caught individuals and are thus very inbred, it is hardly surprising that the analysis of samples from the population of New Guinea highland wild dogs by Surbakti and her colleagues showed close similarity to dingoes but greater genomic variability. The Y-chromosome data suggest that the founder population of dingoes arrived in mainland Australia more than 8,000 years ago and yet remained paleontologically and archaeologically invisible until about 3,500 years ago. It seems improbable that dingoes would not turn up in Aboriginal peoples' living sites or burials during this time span unless, in the early millennia of their occupation of Australia, dingoes did not live intimately with Indigenous Australians as domesticates.[10]

If, as is currently believed on archaeological (not genetic) evidence, dingoes did not arrive in Australia until about 3,500 years before present, then they did not arrive with or participate in the initial peopling of the supercontinent. Thus, my hypothesis that hunting with dogs was an important factor in the survival of early modern humans in Eurasia is not paralleled by events in the peopling of Australia. Australia is distinct in this respect, as in so many other ways.

There is a grave problem in using dingoes as a proxy for human occupation: there is no consensus on what exactly a dingo is. We know with some certainty that dingoes did not enter Greater Australia with the First Australians, but with some later, possibly transient group of people. Yet there is much disagreement about whether dingoes are domesticated dogs gone feral (*Canis lupus dingo* or *Canis familiaris dingo*), as argued by Zhang, or a separate wild canid species (*Canis dingo*), as argued by Matisoo-Smith. Dingoes certainly differ from domestic dogs of similar body size in Eurasia in reproductive pattern, development, vocalization, skeletal proportions, climbing ability, and social behavior and bonding with humans.[11]

Finally, there is little compelling evidence about where dingoes came from because the samples in genetic studies may or may not include pure dingoes. Some scholars estimate that hybrids of dingoes and dogs constitute 78 percent of the feral canid population of Australia. The pure dingo may actually be extinct, although recent specimens from some areas seem to carry no trace of hybridization.

Whenever dingoes arrived, the most probable scenario is that they came to mainland Australia with people by boat. So if we want to understand the migration of humans into Greater Australia, we also need to take a close look at dingoes and the possible parallels between the modern human peopling of the supercontinent and the later dingo invasion.

THIRTEEN

How Invasion Works

IF THE DOMESTICATION OF DOGS
offered a telling advantage to modern humans as they migrated
into, adapted to, and survived in Eurasia, why don't dogs feature prominently in the peopling of Australia? Why didn't the
First Australians bring their dogs with them? The answer is
obvious: when modern humans were expanding out of Africa
toward Europe and Asia, there were no dogs. By the time dingoes showed up, the First Australians had been on the Australian continent for perhaps 30,000 to 60,000 years. This distinction seems key because the adaptations and knowledge that
both humans and canids needed to survive in Greater Australia
differed markedly from those needed by future Eurasians.

The ecosystem to which the First Australians had to adapt
was completely novel and unparalleled in other regions. When
modern humans first appeared in Europe, the basic Ice Age
fauna of Europe broadly resembled the African one where
humans had first evolved; not so in Australia. As humans
moved out of Africa into Eurasia, they also faced archaic

humans who had been living as hunters in that region successfully for hundreds of thousands of years; the First Australians
did not. Thus, comparing the challenges posed by the modern
human migration and adaptation into Eurasia with that into
Greater Australia highlights two crucially important factors.

First, when modern humans arrived in Eurasia, the two-
legged, fire-using, tool-making predatory niche was already
filled by another species, be it Neanderthals, Denisovans, or
some other archaic hominin. Competition within the mammalian predatory guild was almost certainly great. Any improvement in hunting success, such as an imperfect collaboration
with even semidomesticated dogs, would have had a major effect on the success of modern humans.

Second, the indigenous prey species of Australia were naïve
and had not learned to fear humans. This naïveté gave human
hunters a significant advantage. Some of the available prey were
medium- to large-bodied species that used bounding speed as
a major defense against predators. The common Old World prey
species—bovids, equids, cervids, rhinocerotids, proboscideans,
and so on—that galloped, stotted, and pronked away from
predators were unlike the large Australian marsupial mammals
that hopped or bounded to escape. Smaller mammals also took
to the trees or to underground burrows. The marsupial mammals of Australia ranged in size from small, mouse-sized species to very large, slow herbivores such as *Diprotodon optatum*
or *Palorchestes azael* (which had a tapir-like short trunk), and
the rhinoceros-sized *Zygomaturus trilobus*. These latter beasts
might be considered the rough equivalent of proboscideans
or rhinos. The large, ground-dwelling birds with fearsome
claws—cassowaries, the emu *Dromaius novaehollandiae*, the
enormous *Genyornis*—appear to have been swift runners; they
were also highly capable of self-defense with their large clawed

feet. All in all, the distinctness of the Australian fauna would have required new hunting tactics and new knowledge of species' habits. Neither dingoes nor true dogs appeared in Greater Australia until a few millennia ago at most. This fact rules them out as companions or hunting aids to the first humans to inhabit Sahul, yet what we do see in the record is striking.

The arrival of dingoes between roughly 5,000 and 3,000 years ago is indicated by both genetic estimates and fossil remains from human occupation sites, although the estimates of arrival dates do not jibe exactly. If dingoes were not at all domesticated before arriving in Greater Australia they might not have associated with humans immediately after landfall, but that is what the record suggests. The oldest known dingo remains in Greater Australia were found in what appears to be a burial site. This fact might account for the discrepancy between the molecularly estimated time of dingo divergence from the wolf lineage at 5,000 or more years ago and their appearance in the fossil record at 3,250 calibrated years BP (before present; radiocarbon years must be calibrated because the amount of carbon in the atmosphere varied slightly through time). Alternatively, if dingoes were domesticated, or at least tolerant of humans, at landfall, they may have sought out anthropogenic habitats immediately as they looked for other resources.

Currently the oldest known dingo specimen—judged to be a burial—is from the Madura Cave, an occupation site on the Nullarbor Plains dated to 3,450±95 calibrated years BP. The bones from two individual dingoes recovered from this site have recently been directly dated to 3,348 and 3,081 calibrated years BP, giving the oldest reliable date for dingo arrival in the south at 3,250 calibrated years BP. How long before 3,250 years BP was dingo landfall? Jane Balme and colleagues suggest a short time span, noting that domestic dogs with people spread throughout Tasmania in less than 30 years. Klim Gollan estimated continental spread within 500 years of arrival, while Glen Saunders and colleagues estimated 100 years. By comparison, cats dispersed across mainland Australia in about 70 years. None of these estimates would place dingo landfall at 5,000–18,300 or more years ago, as genetic estimates do. Unfortunately, genetic estimates do not actually date anything; they tell you how many years the number of mutations you observe might take to occur. But all genes do not mutate at the same rate, so the number of mutations does not always measure the time since divergence of two forms accurately. As with humans, the most probable place of dingo arrival is in northern Greater Australia, either through New Guinea or Arnhem Land, so getting to the Nullarbor on the south coast of mainland Australia necessitated a very rapid spread through thousands of miles of different habitats. That the dingo spread far and fast suggests it had developed an effective adaptation to the Australian habitats and that its spread may have been facilitated by humans, who created resources not otherwise available to dingoes. The fact that many sites yielding dingo bones include archaeological materials suggests strongly that dingoes were partially domesticated, or at least tolerant of humans, upon their arrival in Australia.[1]

Another major problem in reconstructing the story of the dingo's arrival in Greater Australia is that it is almost impossible to tell by sight if an animal is a pure dingo or one with domestic dog in its ancestry. Even experienced wildlife experts have difficulty classifying animals on sight. Detailed and standardized measurements of the skulls of dingoes, once used to distinguish pure dingoes from dingo–dog hybrids, are no longer considered accurate because known first-generation hybrids show more dingolike morphology than doglike appearance. The difficulty of reliably distinguishing wolves from dogs using most morphometric methods suggests that telling dingoes from dogs will be equally difficult. Genomic comparisons of dingoes and various domestic dogs were initially very promising as a means of identifying a true dingo from a hybrid, but such research cannot be done without capturing or killing the animal, which makes it an unsuitable method for conservation projects. The late Alan Wilton and others discovered a microsatellite—a repetitive bit of DNA that does not code for anything but does get inherited—thought to be unique to dingoes. Additional genomic analyses uncovered a distinct haplotype on the Y chromosome of dingoes and their close relatives, New Guinea singing dogs. But genetic surveys of dingoes' whole mitochondrial DNA (mtDNA) genomes by Kylie Cairns's group undercut previous findings, made by teams led by Peter Savolainen and Arman Ardalan, that concluded there are only two major dingo haplogroup lineages; in total, Cairns and her group discovered twenty different haplotypes among their samples, a shocking difference from two. How had so much variability been overlooked in earlier studies? The key is that Cairns's study examined whole genomes, not samples of a few hundred base pairs.[2]

The two predominant haplotypes found by Cairns's group consisted of one recovered from animals in the north and west and one found in dingoes in the southeast. Although all samples assayed came from "wild" dingoes, the difficulty in telling pure dingoes from those admixed with domestic dogs means that the conclusion that these are dingo-specific haplotypes is open to question. Unhappily, New Guinea singing dogs are also both poorly defined and relatively unknown. All New Guinea singing dogs in captivity are descended from seven or eight individuals captured in the nineteenth and mid-twentieth centuries, so they are highly inbred. They look quite a lot like dingoes and do not bark, but "sing" (or howl) in chorus. Their coats are often ginger-colored or red, but they may also be various shades of tan or brown with white on the underside of the chin, on the paws, and on the tail tip. They have fluffy tails and a double coat; relatively broad, wedge-shaped heads; and shorter legs than dingoes. Their ears are naturally erect. They weigh about 8–10 kilograms as adults.

The high-quality samples from three individuals from the wild population of New Guinea highland wild dogs have now been analyzed and are affirmed to be the original population from which New Guinea singing dogs were captured. These wild canids are closely related to the inbred captive population of New Guinea singing dogs, slightly increasing the known genetic variability of that canid. We now know with reasonable certainty that these New Guinea highland wild dogs represent the surviving founder population of New Guinea singing dogs and that they are closely related to pure dingoes from Australia. Haplogroups (genetic groupings inherited from a single ancestral lineage) on the Y chromosome were analyzed by Benjamin Sacks of the University of California, Los Angeles, and his colleagues, who were sampling what were thought to be unadmixed male dingoes. Their analysis also suggests the presence of two lineages. The northwestern dingoes share haplogroups H60 and H3, which makes them closer genetically to the New Guinea singing dogs in captivity, whereas the southwestern dingoes had H3 and H1 haplogroups. However, these Y-chromosome data conflict with those from the maternally carried mtDNA among dingoes, which suggests that the female dingoes from the southeast are more closely related to the New Guinea singing dogs. This apparent conflict may reflect behavioral flexibility among dingoes of different sexes.

Perhaps it is easier for females to be accepted into new groups than it is for males, or perhaps there is a problem with the sampled animals. The real catch is that we do not know if the dingoes in these and other studies were actually full dingoes and not hybrids. Ardalan and colleagues explain: "Blood samples were collected from 47 captive and wild unrelated male dingoes at different locations in Australia. . . . Effort was taken to sample dingoes with as little admixture with domestic dogs as possible based on analyses of dingo diagnostic microsatellites as well as phenotype." It is not clear how Ardalan and colleagues knew that the wild animals they sampled were unrelated to others, though this is a reasonable assumption for animals collected at considerable distances from others. Until a recently funded genomic study of the earliest, precolonization dingo specimens in Australia is completed by Melanie Fillios—using specimens derived from mummified remains, skins, fossils, and remains in museums—we will not know how to identify a dingo properly even from a complete genome. Until that information is gathered, the results of all of the genomic studies must be considered provisional. Such a study is a key priority in understanding the human–dingo relationship.[3]

Comparing the First Australians' adaptations to Australia with those of the later-arriving dingoes' adaptations promises to clarify the challenges that were faced by placental predators that migrated into the continent. As predators in different niches but the same environment, the two very different species showed somewhat similar patterns in group size and composition, in adaptations to a cold and dry environment, and in use of available resources.[4] As the First Australians did, dingoes live in fairly small groups comprising extended families. And as it was to the First Australians, knowledge of water sources is important to dingoes. Though it is difficult to measure dingoes' knowledge of water sources in their territory, it has been shown that, physiologically, dingoes have evolved to withstand longer periods of time without access to fresh water than most canids. They are believed to be able to detect underground water and dig out soaks that benefit many species. One observational study of radio-collared dingoes in the wild reported that some individuals went as long as twenty-two days without going to a water source. Similarly, Indigenous Australians in arid regions were apparently adapted to withstand extreme nighttime cold and dehydration with less severe consequences than groups of European humans under the same conditions.[5]

Nomadism was also a key feature of the human adaptation to Australia. It is clearly possible to "eat out" the game and plant resources in an area, so a nomadic habit is advantageous, even when the landscape is being managed by fire-stick farming (intentional burning), irrigation, and other strategies. In fact, in areas where keeping dingoes is legal in modern Australia, the regulations for keeping them are stringent and reflect a strong tendency for dingoes to escape captivity or confinement rather than staying in one place. Because many Australian ranchers fear that free-ranging dingoes will attack stock animals, pets, and children, preventing dingoes from escaping in areas where they can be kept as pets is important.

A wide-ranging diet, particularly one including the use of coastal food sources, seems to have offered an advantage to the First Australians. Similarly, dingoes eat a broad spectrum of prey (including fish) and plant foods according to the season and location. Though dingoes may concentrate on one abundant animal in a particular area, when they move on they may exploit a very different resource. The opportunistic diets of dingoes in the wild indicate that they hunt smaller vertebrates alone or with one companion, and dingoes hunt in packs only when larger animals are abundant.

Though the details vary with habitat and cultural group, dingoes were part of Indigenous Australians' lives. Traditionally, Indigenous Australians frequently kidnapped dingo pups and raised them in their camps. In colonial times, as Jane Balme and Sue O'Connor note, dingoes were ubiquitous and almost invariably associated with humans; it is difficult to find an ethnographic or historical image of mainland Indigenous Australians' camp life or gatherings that does not include dingoes. Dingo pups were treated as pets, as companions, as blankets, and as guardians against human or supernatural beings. Dingo pups were fed, often suckled by Indigenous women, de-fleaed, kissed, and coddled. (Some people from Western countries who have attended presentations I have given were shocked at the thought of women suckling pets or livestock, but this is not an uncommon practice among nonindustrial peoples, who can be quite casual about nourishing animals and humans alike with "leftover" breast milk.) Rather remarkably, dingo pups were carried around women's waists on long journeys or where the footing was difficult for tender puppy feet. Women more often seemed to have special relationships with dingoes than men did and owned, on average, three times as many dingoes as men did. These behaviors suggest that dingoes were reasonably well adapted to anthropogenic habitats. In some regions, dingoes were traditionally used to flush out game; in others, observers found them to be useless as hunters, often frightening large game away. Balme and O'Connor hypothesize persuasively that dingoes often accompanied women who were gathering plants, shellfish, or small game such as goannas, rats, snakes, or possums. The presence of dingoes may have decreased the search time for such species, which they could smell or hear better than humans.[6]

Traditionally, dingoes were also considered to be able to detect evil spirits or ghosts and were highly valued as protectors of women and children. The world is viewed as full of potential dangers in Dreaming lore, whether that danger takes the form of poisonous or venomous creatures, strangers, or ancestors seeking revenge for a failure to comply with prescribed behavior.

Close connections between dingoes and Aboriginal peoples of Australia are well documented in colonial and recent times. On this basis, Balme and O'Connor suggest a testable hypothesis: The combination of information on the present-day dingo diet and the ethnographic and historical evidence suggests that the introduction of the dingo and the close bonds that were formed between them and women in the camps had the effect of increasing the rate at which hunting women encountered game, thereby increasing women's contribution of meat to the diet. If this is the case, we would expect to see a greater variety of small to medium-sized game, and a proportional increase of such animals (at the expense of large game) in archaeological sites over time. Few archaeological sites sample a mid- to late-Holocene timespan and have been analyzed in sufficient detail to test this hypothesis, but those that do exist show precisely the sort of change in prey that Balme and O'Connor predict. More important, dingoes were deeply integrated into traditional knowledge, corroborees, and songs. Because Dreaming Law is acknowledged to change through time as more truth is revealed, it has been suggested that when dingoes arrived they assumed preexisting, mythic roles that had been previously assigned to the marsupial thylacines. It has been suggested that dingoes were instrumental in the extinction of thylacines, which were rather doglike in appearance but had an enormous gape and boldly striped backs and tails. Many myths identify dingoes as the ancestors of humans, who, in the Dreamtime, taught humans how to behave. "Mother and Father Dingo Make Aboriginal" is a translation of part of a Yilngayari creation myth.[7]

That ancestors could be both human and dingo, alternately or even simultaneously, became a fundamental premise of Indigenous Australian beliefs. This duality of identity confounds many Europeans, yet it speaks directly to the special status of dingoes in Indigenous Australian cultures. Merryl Parker researched dingoes' dual roles in Indigenous Australians' and White Australians' mythologies. In the Indigenous Australian myths and traditional stories, dingoes are not only human ancestors, but can also turn into humans and back again.

The burial of canids in ways similar to those used for humans is seen by many archaeologists as a clear indication of the domestication status accorded to the canids. Sometimes, dingoes were (or are) buried the way people are. By way of example, Ian Cahir and Fred Clark cite Samuel Rawson, a squatter living southeast of Melbourne who in 1839 wrote of how he had shot some of the Boonwurrungs' dogs (which at that date might have been pure dingoes) for killing his poultry. Rawson noted in his diary the reaction by the dogs' Indigenous owners: they buried the dead bodies of their four-legged companions with great ceremony, wrapping them in blankets and sheets of bark and lighting fires by their graves—after which they decamped and moved up the river. Cahir and Clark also quoted the settler William Thomas, who recorded that Victorian-era Indigenous people performed mortuary ceremonies for their dogs: "The

Sometimes dingoes were buried in a manner similar to humans. Ben Gunn discovered a dingo burial in a rock shelter in Arnhem Land during a survey for rock art (*top*). The dingo was wrapped intact in bark cloth, and the bundle was deposited on a shelf and protected from disturbance with a ring of rocks (*bottom*). The bones were not ochred. Jawoyn elders told Gunn that, to their knowledge, this was an unusual treatment. Direct dating of the bones suggested that the dingo died (and was buried) between 1680 CE and 1930 CE.

Blacks have various kinds of Ngar-gees (or Corroberry's) including a ceremony in burying their dogs. No other animal is regularly buried by traditional Aborigines."[8]

In 2010, Ben Gunn and colleagues described a dingo burial that he and his team discovered in Arnhem Land during a survey for rock art. It was located in an area of the central plateau with many rock shelters that had an unusually dense concentration of rock art. The rock shelter in which the burial was located had faint white paintings of dancing figures with headdresses imposed over a red kangaroo and beeswax designs on the walls of the shelter. The dingo's entire body had been wrapped in bark cloth and placed on a high ledge within the rock shelter soon after death; the bones of the skeleton were still articulated. Several rocks were stacked up to prevent disturbance of the carcass and three partly burned logs were also placed on the ledge, which might have been to protect the bundle or could simply have been firewood stored for future use. Conventional radiocarbon dating of a vertebra and a rib bone revealed a high (95.4 percent) probability that the dingo died between 1680 and 1930 CE.[9]

Human burials by the Jawoyn, on whose lands the dingo burial was discovered, have many similar features, but elders told Gunn and his colleagues that burying dingoes was unusual:

> For the Jawoyn, a traditional human burial involved two stages: an initial tree burial of the body, followed by its retrieval some months later when the bones had been desiccated. The skull and long bones were then ochred, wrapped in paperbark, and placed in a rockshelter within the clan territory. . . . The bundles are commonly placed in a small cleft or on a ledge in

> the shelter and then protected by a rock surround. . . .
> Over time, animals may disturb the burials, at which
> time relatives or other visitors to the shelter may place
> the skull prominently on a ledge facing out over their
> country as a sign of remembrance.[10]

Though dingoes are not prominent figures in the Jawoyn mythology of the present, they do play major roles in the mythology of other Arnhem Land groups.

Another canid burial from Jawoyn country, with similar
treatment and placement of the bark cloth–wrapped bundle,
was discovered by Gunn and his colleagues later in that same
archaeological art survey, about forty kilometers south of the
first burial. The second canid was provisionally judged to be
more probably a dog or a dingo–dog hybrid, but the identity
of the animal is unclear because the bones could not be removed from the burial. Conventional radiocarbon dates of
this canid suggest the date of death was between 88 ± 25 years
BP, making it effectively contemporaneous with the first.
Modern Jawoyn elders asserted that the practice of burying
dingoes or dogs was unusual and possibly the work of a single
individual to pay tribute to a beloved pet. Dingo burials are
also known from other areas, including South Arnhem Land,
Queensland, Wardaman country west of Katherine, and in the
Keep River area near the western border; they occur sometimes in association with rock art or human burials. These
burials confirm the observation that dingoes acquired a status
as spirits or supernatural figures in the eyes of traditional Australians. The remains of dingoes, and of dingo burials, suggest
that dingoes acquired near-human status soon after arriving
in Australia, whether or not they were truly domesticated. In-
terestingly, when true domestic dogs arrived during the colo-

nial invasion of Australia, they were taken up by Indigenous Australians almost immediately. They were highly desired and admired as hunting aids and companions. But neither dogs nor dingoes were available to the First Australians.[11]

The common scientific assumption that the first dingoes arrived with the First Australians was overturned in 1960 by the discovery of a fossilized dingo skeleton at Fromm's Landing that was dated to 3,000 years ago. Among the Aboriginal peoples' accounts about the dingo that Merryl Parker uncovered in her research was a corroboree still performed in 2000–2005 by the Kundi-Djumindu people from the coast of the Northern Territory, in which the dancers portray the dingoes running excitedly along the deck of a boat, jumping into the water, and swimming to shore. In watching a video of such a performance, I personally found the dancers portraying dingoes to be readily recognizable as such. The boatmen were visitors from some other land. The question arises: Who are, or were, those visitors who brought dingoes? In separate works, Parker and Melanie Fillios and Paul Taçon suggest that they may have been Makassan people who came to collect sea cucumbers for sale to Chinese traders, even though there is no firm evidence that the Makassan traders began voyages to Australia earlier than about 1500 BCE.[12]

Parker found fifty published myths about dingoes, though to some extent they were reshaped by the English speakers who wrote them down. Many Europeans have found Indigenous Australians' myths rather cryptic in that they are very brief and assume a good deal of background knowledge on the part of the listener that Europeans lack, so explanations are added when they are written down. These myths often explain the creation of water sources or unusual rock formations. Thus, a key question is how dingoes that have been in

Australia only a few thousand years could possibly have shaped the landscape, dug rivers, or left huge boulders as reminders of dingo pups—but time, in Indigenous Australian beliefs, is not necessarily linear. Along the same lines, Roland Breckwold suggested that dingoes may have replaced thylacines in traditional stories as the former became more common and the latter rarer and eventually extinct. Though the Law expressed in Dreaming myths is regarded as unchanging through time, adjusting the exact language to substitute dingoes for thylacines, or dogs for dingoes, is entirely consistent with Indigenous understanding.[13]

Darcy Morey and others who work in Europe or the Americas maintain that being buried as if they were human is one of the most irrefutable signs that canids have attained a near-human status. There are numerous burials and even recognizable cemeteries for presumably domesticated dogs in Eurasia starting about 14,000 years ago, but the alliance between dingoes and humans is not manifest in this way in Greater Australia until very much later in time. Though I have suggested that cooperation between humans and canids was a vital advantage to the modern humans who invaded central and eastern Europe and Asia and who faced both other hominins and an Old World fauna, this was not the case in Australia. Humans and canids simply did not co-occur (much less live commensally) in Greater Australia until thousands of years after human landfall. Canids cannot be credited with human survival over competitors in Australia the way they can be in Eurasia. Why, then, did the First Australians succeed?

The First Australians did have some notable advantages over early modern humans in Eurasia. One was that the First Australians encountered neither many large predators nor any other hominins in Greater Australia. Also, because the mammalian predatory guild in Sahul was less diverse and composed of species that were physically smaller than those in Eurasia, the First Australians probably faced less intense competition. There were only two medium-large mammalian, marsupial predators at the time of human arrival in Greater Australia: the thylacine and the marsupial lion. However, it would be unwise to overlook the potential competition from formidable reptilian or avian predators in Sahul. Another advantage was that, as maritime peoples, the First Australians knew how to extract marine food resources; these were probably the basis of their diet, supplemented with terrestrial resources. This dual economy was apparently very viable. Finally, the terrestrial fauna, which the First Australians learned to exploit fairly quickly, was naïve to hominin hunters.[14]

The novelty of the resources available to the First Australians must also have been a key issue. When modern humans arrived, Australia was hot, many parts were desertic or seasonally arid, and the country was inhabited by unfamiliar animals. The plants—some of which were poisonous unless processed—were new, and rainfall and water flow were unpredictable. Not only the fauna but the flora and the very landscape itself were outside the knowledge that the First Australians had gained elsewhere. Little of the specific, detailed knowledge about plants and animals that enabled humans to survive in Africa, southern Asia, or the Levant was of immediate usefulness in Greater Australia. However, I speculate that the gathering of such knowledge was understood to be very important, an idea that has also been expressed by others. The importance of such knowledge is manifest in the early development of symbols and artwork that communicated key data about the new ecosystem through a network of connections. At some point that is as yet impossible to date securely,

Indigenous Australians also developed a system of songlines, oral traditions, symbolic depictions, and dances that encoded this precious information (and much more) in ways that could be and were handed down for millennia. For nonliterate people, such a system could be a vital means of recording and storing information.[15]

Based on an analysis of oral traditions of twenty-one coastal Indigenous Australian groups, Patrick Nunn and Nicholas Reid argue that memories of the inundation of the Australian coast were retained for at least 7,000 years—an astonishingly long time, and an antiquity that is much greater than previously suggested by folklorists. Indonesian rock art predates the earliest Australian archaeological sites, as does the use of ochre and the manufacture or wearing of ornaments and artifacts made from shells, eggshell, bone, or teeth in African sites. This use of symbols and personal adornments is among the attributes often seen as signs of fully modern cognitive abilities. Though oral traditions are even more difficult to date than rock art, there are some important indications that such traditions have considerable antiquity. This extraordinary preservation of information that could be lifesaving reaches much farther back into history than oral historians have generally believed is possible.[16]

As emphasized by Nunn and Reid, memories and knowledge are most likely to survive and be remembered under three conditions. First, the knowledge is held in relatively isolated cultures that are not exposed to others. Second, the information is regarded as being of great importance to survival, and that importance results in socially effective, formalized mechanisms for its transmission to the next generation, with particular people held responsible for teaching youngsters. To learn and teach such stories as the coastal people's stories of

inundation was the task of unmarried men, and their performance was judged by potential mothers-in-law. Failure to
teach the important information could lead a parent to refuse
to let a young man marry a daughter of their family even if
she had already been promised to him, so the stakes were high.
Third, the knowledge is tied to aspects of the physical landscape, which is obviously true in traditional Aboriginal peoples'
cultures. I would add to these criteria a fourth: that the possibility of long-term remembrance of information is enhanced
when it is transmitted in multiple media, such as recital, stories, song, dance, and art.

Both elaborate systems for retention and transmission of
knowledge and of the necessary information for exploiting
uniquely Australian resources show clearly how critical gaining
and retaining knowledge must have been to the First Australians. New knowledge that was amassed, codified, and shared
was absolutely essential to their survival, and the knowledge
is still told and retold as a living truth.

Ethnographic evidence suggests that the Dreaming is not a
fixed canon but a constantly evolving one as new wisdom is
revealed to those entrusted with it. Dingoes appear in many
traditional stories that teach moral values and behaviors. For
a nonliterate people, the Dreaming also recorded and preserved
vital new information specific to Greater Australia—the locations of both ephemeral and permanent water sources, for example. That vital information also included which vegetable
foods were edible, which required processing before they
could be eaten, where and when particular foods could be
found, and where to hunt the indigenous prey species.[17]

Enhancing the First Australians' detailed botanical knowledge was an understanding of the timing and intensity of fires
that would favor regeneration of different flora and also create

ecological zones that would attract different fauna. We cannot assign a date to the discovery of these practices, and the amount of charcoal in dated pollen samples as yet yields no definitive answers. But the simple fact that Indigenous Australians survived to the time of the first European landing—and indeed, to the present day—attests to their success in adapting to the new conditions offered by the new continent.

Unlike the first European settlers, the First Australians had no eagerly awaited ships from "home" bringing supplies to keep them alive. There were no sheep or cattle or seeds available as foundation stock for vast agricultural estates. The First Australians survived because they assessed the ecosystem and its exploitable resources accurately and passed that vital information on to others, not because they were well supported by a mother country. The invasive European settlers not only lacked the vital information amassed by Aboriginal peoples over millennia, but also had no formal means for retaining and sharing what knowledge they gained.

New dating using state-of-the-art techniques has established that both Tasmanian devils and thylacines became extinct on the Australian mainland within a very brief period between 3,227 and 3,179 years BP, dates that technically must be considered synchronous. Further, this period falls fairly shortly after the earliest recorded presence of dingoes on the mainland at about 3,250 calibrated years BP.[18]

Given the ecomorphological similarity of dingoes and thylacines, the arrival of the placental dingo certainly brought it into competition with thylacines. Species from elsewhere, such as the dingo, predictably put pressure on the native species with the most similar role in the ecosystem, in this case the thylacine. An important issue in determining the outcome of such encounters is which species would most often dominate in direct interference competition, such as trying to steal another predator's kill. Dominance is often correlated with body size, though pack hunters tend to dominate similar-sized solo hunters. When Melanie Fillios, Matthew Crowther, and Michael Letnic embarked on a detailed study of thylacine and dingo size to evaluate the competition between thylacines and dingoes, they found that male Tasmanian thylacines were bigger than dingoes, yet male mainland thylacines were comparable in size to dingoes. Female thylacines in either location, Tasmania or the mainland, were significantly smaller than male thylacines or dingoes. As for hunting habits, dingoes certainly live and hunt in family groups, though such behavior is more common in them than in thylacines, according to extant documentation. Dingoes also have a larger brain per unit body size and a higher metabolic rate than thylacines, meaning they needed to eat more to survive. Fillios and colleagues describe dingoes as "smarter and hungrier" than thylacines. Therefore, they argue that, on the mainland, dingoes would likely have dominated direct interference encounters with thylacines. Preferential killing of smaller, female thylacines by dingoes could also plausibly have depressed the reproductive output of this species, favoring extinction. Because dingoes never reached Tasmania, their absence might account for the larger size of Tasmanian male thylacines and the late survival of all thylacines on Tasmania. This interpretation is fascinating but still somewhat speculative.[19]

In another paper, Fillios and colleagues demonstrated that archaeological sites that preserved information about prey taken by humans both before and after the introduction of dingoes demonstrated that humans shifted to smaller prey over time. It seems likely that this reflected a greater scarcity of larger prey once dingoes were introduced.[20]

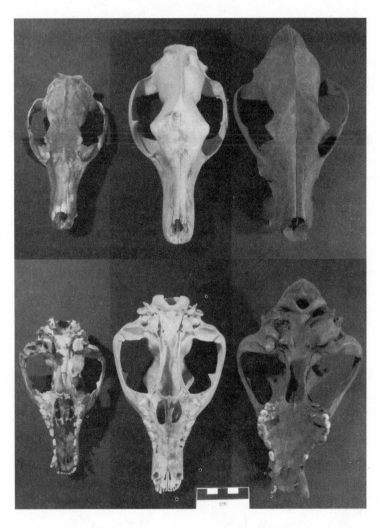

One of the questions persisting in Australia is whether the appearance of dingoes had anything to do with the extinction of mainland thylacines. This photo shows the skulls of a female thylacine (*left*), a male thylacine (*center*), and a male dingo (*right*) from mainland Australia. Female thylacines were smaller than males, but mainland dingoes were as large as or larger than either. Some scholars have suggested that preferential killing of females by dingoes may have pushed mainland thylacines into extinction. Each block in the scale is 1 cm.

What we see, then, is that dingoes probably entered Australia as an undomesticated or incompletely domesticated species that nonetheless had some tolerance for humans and their settlements. Humans had already adapted well to the new continent and its unique challenges by the time dingoes arrived. The First Australians had already survived, invented some new tools and mechanisms for thriving, and expanded their well-known territories into different ecosystems. Humans had no need for canine hunting assistants, nor did dingoes need humans in order to hunt effectively. In fact, there was certainly greater competition for large prey once dingoes entered the ecosystem. However, a thorough search of the ethnographic literature by Loukas Koungoulos and Fillios revealed that dingoes were sometimes used to hunt larger prey, such as kangaroos and emus, and to drive them into pits, nets, or simply heavily vegetated areas where it was easier for people to hide, wait, and then dispatch the prey. If dingoes or their ancestors had been domesticated before arriving in Australia, dingoes may indeed have become feral, or lost some of their domesticated traits, soon after arrival in Greater Australia and yet have affected the ecosystem markedly.[21]

Still another perspective on the adaptations of species new to Australia can be seen by looking at the other placental mammals introduced during colonial times. Europeans brought in horses (now running wild as brumbies), cattle, sheep, camels, cats, foxes, and rabbits, to name the most common imports. Horses of various breeds were brought in with the First Fleet of settlers and convicts in 1788, for pulling plows, transport, and mustering stock animals. Brumbies are hardy and surefooted and survive well in the outback. They also reproduce rapidly in favorable conditions, having no natural predators, and compete with cattle or sheep for grass. Their tough hooves compact and degrade the soil, as in moss bogs, and may degrade riverine habitats. Deliberate culling of brumbies is practiced but highly controversial in Australia.

Cattle and sheep were imported as livestock, particularly those breeds that adapt well to dry conditions. They are largely stacked on stations where they forage for themselves until rounded up. There is some evidence that the presence of dingoes favors stock animals because dingoes work to lower the number of kangaroos and emus that compete with sheep for grass on the exclusion side of the dingo fence. The dingo, though not an indigenous species, appears to interact with, and possibly regulate, other invasive or native species that compete with sheep for grass.[22]

Both domestic and wild European rabbits arrived with the First Fleet and with subsequent immigrants; they were intended to serve as game animals. The rabbits bred explosively; rabbit plagues posed a major problem in Tasmania by 1827 and in mainland Australia by 1866, decimating crops. Various measures to control the rabbit population were tried, including shooting, the use of poisons, the deliberate introduction of myxomatosis and other fatal rabbit diseases, building of a rabbit-proof fence in Western Australia in 1907, and the use of ferrets for hunting. None of these methods has solved the problem.

Dromedary camels were imported from British India and Afghanistan for use in desert transport and freight haulage in the arid outback, particularly in the central and western states and the Northern Territory. For example, camels and camel handlers were specifically imported from India for use in various expeditions, including that by explorers Burke and Wills. Camels survive well in Australia when turned loose to forage for themselves; the estimated feral population is now more than a million animals. Culling by gunshot from helicopters has been used to try to control camel numbers in areas where drought and bushfires have led camels to destroy fences, gardens, and water sources.

Like dogs, cats were brought in as pets in the colonial years; millions are now feral and roam the landscape freely. They are thought to be responsible for the extinction of many of the smaller Australian marsupial species, and efforts to eliminate feral cats altogether continue. Another imported predator, the red fox, is similarly destructive to small marsupials. Red foxes were imported to enable fox-hunting in Australia and Tasmania, though they did not survive to form a feral, self-sustaining population in Tasmania. It is thought that foxes declined in Tasmania because of the competition of Tasmanian devils. In the hope that importing Tasmanian devils to mainland Australia might regulate fox numbers, twenty-six Tasmanian devils have been released into a protected sanctuary by Aussie Ark, Global Wildlife Conservation, and WildArk. If this initial group survives well and avoids the contagious facial tumor disease that has killed many Tasmanian devils, more will be released in hopes of restoring a natural regulatory agent in Australia. Foxes now infest much of Australia proper, and poison and hunting are the primary methods used to control the species. There is growing evidence that the presence of dingoes helps suppress the populations of cats and foxes but does not completely control them.[23]

The overall conclusion to this brief review of the impact of placental species in Australia is that they have often proven deleterious to the native fauna through both competition and outright predation. The Australian fauna is well adapted to its unique and difficult circumstances, but there are few indigenous predators and, in general, a lower population density of species than on other continents.

FOURTEEN

A Different Story

AS THE FIRST AUSTRALIANS wended
their way unknowingly to Greater Australia, other populations
moved north and east through China, Tibet, Mongolia, Russia,
and Siberia and, eventually, to the Americas. The Eurasian
group faced archaic humans and a much denser and more
diverse group of predators with which they had to compete.
Before humans were able to invade the last continent—the
Americas—many millennia later, they first had to adapt to the
cold climate of Ice Age Eurasia.

One of the most important sites to yield information about
the modern humans' invasion of the far north is the Yana
Rhinoceros Horn site in Siberia, where humans lived about
33,000 years ago. The site includes many stone tools and a
truly astonishing array of faunal remains that were altered by
humans, including mammoth bones with fragments of stone
tools still embedded in them; vessels crafted out of ivory and
decorated with engraved geometric and possibly anthropomor-

phic designs; beads, needles, and awls made of bone, tooth, and ivory; and remains of large animals processed for eating. Most of the cultural remains were made of materials that are rarely preserved in such detail, making for a stunning assemblage. The tools were skillfully and bifacially shaped, a quality of manufacturing seen as typical in the later Clovis tools. These wonderful artifacts show the sophistication of the skills, mythology, and perhaps shamanistic beliefs of the people who moved into Beringia, which connected what is now northeastern Asia to northwestern continental America. Whether the descendants of those same people later migrated south, possibly through an ice-free corridor, or adapted to a maritime existence that led them southward via a coastal "kelp highway," the people of the Yana Rhinoceros Horn site were fully adapted to the harsh Siberian conditions and the Ice Age animals and plants of the far northern Arctic region. What this wealth of information leaves unresolved is where exactly the Yana Rhinoceros Horn site people came from and when they entered the Americas.[1]

The third human continental expansion—into the Americas—was much, much later in time than the migrations into Eurasia or Australia. Current evidence suggests that one population from Siberia migrated across a land bridge into the landmass known as Beringia about 21,000 years ago. Many scholars believe human populations spent thousands of years there—in the so-called Beringian standstill—genetically isolated from later migrants from Siberia. (An important fellow inhabitant of Beringia was probably the gray wolf.) Later, some of these humans moved from Beringia into the lower Americas and gave rise to the main group of Indigenous Americans. This scenario is largely supported by recent genetic studies indicating that the humans who invaded the Americas split into two large groups,

One of the most spectacular sites in Siberia is the Yana Rhinoceros Horn site. The Arctic climate has preserved numerous artifacts of stone, bone, wood, antler, and ivory dating to 33,000–28,000 years before present, including symbolic, sophisticated, and highly decorated vessels such as the one shown in this photo.

one of which is related to the Native Americans who lived at the time of contact with Europeans and one of which apparently stayed in Beringia and largely died out. As with the First Australians, there is much debate about how humans first reached and then spread throughout the Americas.[2]

When I was a graduate student, the "Clovis first" theory held sway. According to this scenario, people came across

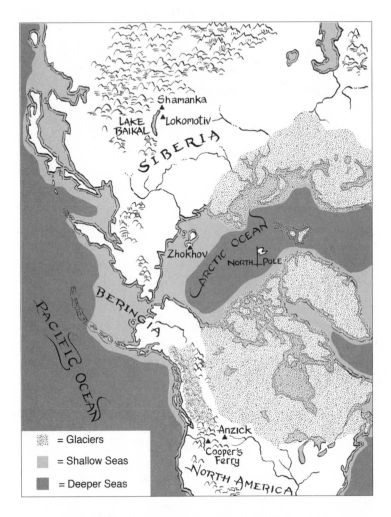

This map shows some of the important Siberian archaeological sites, the
North Pole, Beringia, and the ice-free corridor from Beringia that ran down
between the Cordilleran and Laurentide glaciers into North America. The
ice-free corridor opened about 15,000 years ago. Humans may also have
used the so-called kelp highway along the shallow seas of the Pacific Coast
as a route to North America.

Beringia about 13,500 years ago, traveling down into North America in an ice-free corridor between the Cordilleran and Laurentide ice sheets and leaving behind tools made in a distinctive style. An impressive site featuring a cache of large tools made in this style—and also featuring dead mammoths, mastodons, and other large prey species—was found in Clovis, New Mexico, which gave this people and culture its name. For many years, no credible sites older than the Clovis sites and no tools of a different and older style were found. It did not hurt the acceptance of "Clovis first" that the tools are quite gorgeous. They sometimes are found in what appear to be special caches or are associated with the apparent hunting of mammoths and mastodons—heroic-sized animals. Archaeologists who claimed to have found pre-Clovis sites were often harshly criticized and their claims greeted with deeply rooted skepticism. At the time, as a student with little familiarity with stone tool technology, I nonetheless suspected that the opinion of my elders and betters was too dogmatic, though I had no data to support this suspicion. If it were decreed as if by fiat that no sites could possibly be older than 12,500 years old, it seemed to me that such localities would not be found, even if they existed, because nobody would look in older sediments.

The exceptional dominance (and consequences) of the "Clovis first" paradigm on American archaeology until 1997 should not be underestimated. This paradigm collapsed with an unprecedented site visit by scholars, not only to institutions that held key artifacts but also to Chile to visit and inspect Monte Verde, a site that was excavated for decades by Tom Dillehay of Vanderbilt University and his colleagues. This site had exquisite preservation of stone tools, bones, hide, hearths, and plant remains. The participants in the site visit were a group of well-respected scholars with research expertise in

Part of the reason the "Clovis first" theory persisted for so long was the beauty of Clovis points and the obvious skill used to make them.

various aspects of Paleoindian sites and cultures, but it was a mixed group, some of whom held "Clovis first" views and some of whom did not. After a week of viewing artifacts, data, presentations, and the site itself, the views initially expressed by the panel had altered. The main area of the Monte Verde I site was judged, unanimously, to be archaeological in origin, with a firm date of 14,500 years before present. A secondary, deeper occupation level dated to 33,000 years was judged to be ambiguous still. The report by the panel published in *American Antiquity* is remarkable; it concluded:

> The Monte Verde site has profound implications for our understanding of the peopling of the Americas. Given that Monte Verde is located some 16,000 km south of the Bering Land Bridge, the results of the work here imply a fundamentally different history of human colonization of the New World than envisioned by the Clovis-first model and raise intriguing issues of early human adaptations in the Americas.[3]

In other words, the site visit was a public trial by fire in which a skeptical and well-respected panel evaluated the site, the landscape, the artifacts, and the evidence—and demolished the Clovis first paradigm. The peopling of the New World had to be rethought, once even one pre-Clovis site was more widely accepted, with that acceptance based on careful excavation, new techniques, and cautious interpretation. The irony is that the Clovis first paradigm was shattered by a site almost as far south in the Americas as it was possible for anyone to go.[4]

The traditional theory held that humans entered the Americas by traversing an ice-free corridor that served as a path from far eastern Siberia through the exposed landmass of

The site of Monte Verde, Chile, finally convinced archaeologists that there were genuine pre-Clovis sites in the Americas. Among the remarkable artifacts preserved at Monte Verde were the wooden foundations of a long house divided into twelve segments (*top*), each probably used to house a single family, and wooden pegs that anchored residential wood and hide structures (*bottom*).

Beringia and down into North America—except when the route was blocked by the joining of the Laurentide and Cordilleran glaciers. Several archaeologists, including Jon Erlandson, Ruth Gruhn, and Knut Fladmark, proposed that people migrated from Siberia to the Americas by following a coastal-hugging route often called the kelp highway, a route that was rich in resources such as fish, shellfish, and marine mammals. Erlandson wrote, "I believe the very scarcity of pre-Clovis sites is significant—suggesting that we may be missing an important part of the record. Small and highly mobile populations may explain the scarcity of early sites in some regions, but rising post-glacial sea levels and the inundation of vast areas of the continental shelves are also a major problem." The reason, he argued, that sites to support this hypothesis are not abundant is that sea levels rose during the Last Glacial Maximum (from about 26,500 to 19,000 years ago). The warming environment flooded Beringia under many meters of water. The difficulties in discovering sites with evidence of the timing and details of a migration into the American continent by boat when most of the sites are underwater closely parallels problems in investigating the route taken by early modern humans to get to Greater Australia. In both cases, the logic supporting the idea depends on the reliability of maritime resources along the coast and the ease of long-distance transportation once boats, rope, and navigation have been mastered.[5]

Once again, the key problem is the absence of solid, tangible evidence. Are there so few sites revealing how and when humans got into the Americas simply because there are none that are above water? Or are there so few because there never were any such sites because archaeologists are looking in the wrong places at the wrong time?

Some suggest that southern Beringia served as a refugium for ancient North Siberians during the very cold Last Glacial Maximum period (the Ice Age). Until the climate warmed, starting about 16,000 years ago, the human population may have stayed for thousands of years in southern Beringia, stopped and isolated by glaciers that prevented them from moving farther south. This period of isolation (20,000 to 16,000 years ago) ensured that there would be little additional gene flow from outside during the period and permitted the development of some novel Native American haplotypes that are not found in Asia.[6] Then the climate became less severe and glaciers declined, exposing more land and allowing humans to enter the Americas about 15,000 years ago. This so-called Beringian standstill hypothesis was based on genetic studies that identified a scant four major, and three minor, mitochondrial DNA (mtDNA) haplotypes among all Native Americans. At some point, the founding population split into two genetic groups, a more basal northern group and a southern branch, which contain all the genetic diversity known among both ancient and living Native Americans. The small number of haplotypes overall suggests that the number of founders of Native American populations was initially low.

In 2013, discoveries at a site in western Alaska known as Upward Rising Sun, dated to 11,500 years ago, clarified some of these issues. The dating of the site suggests that the humans who lived there might have belonged to a population that included some who had survived the Beringian standstill and some who were among the first to enter North America south of the ice sheets. Remains of two infants excavated at Upward Rising Sun (known as URS1 and URS2) yielded samples that showed they were both female, one about six weeks old and one a late-term fetus. They were deliberately buried together with antler and bone tools, in a circular hearth or pit, and covered with red ochre. The discoverer, Ben Potter of the University of Alaska, Fairbanks, and colleagues were able to recover more intact genetic material from the older girl than from the younger. Though they were buried together in a ritual fashion, the girls were not sisters; each carried distinctive mtDNA, inherited from their mothers. The two girls represented two of the major haplotypes that gave rise to all modern Native Americans, and the older girl had basal mtDNA from the northern branch of Native Americans.[7]

The peopling of the Americas is a messy story. Some of the Beringian people migrated back into Siberia, becoming a group known as Ancient Paleo-Siberians, and still others moved farther eastward into North America, becoming a group known as the First Peoples. Very few remains of these First Peoples have ever been recovered as fossils, and those that have tell a complicated and convoluted story. Among the earliest human remains yet found in the Americas are Kennewick Man, dated at 9,000 before present; Santa Rosa Woman, whose femurs were discovered on Santa Rosa Island off southern California, who lived about 13,000 years ago; and Anzick Boy, discovered in a burial in Montana that dates to 12,707–12,556 years before present and includes more than a hundred tools made of stone and antler. None of these earliest Americans were associated with canids, though later people were. Had they left most dogs behind? Why? It is hard to tell.

Late in 2020, Anders Bergström and his team published a paper on prehistoric dogs and what they might reveal about human (and canid) migrations. The team sequenced the genomes of twenty-seven prehistoric dogs from Europe, the Near

East, North America, Greater Australia, Asia, and Siberia, ranging in age from almost 11,000 years ago to 100 years ago, to examine the relationship between these dogs and the associated human populations. The study confirmed that all dogs shared a common ancestry from ancient wolves, with limited genetic transfer from modern wolves since the time of domestication. By 11,000 years ago, five major dog lineages were established, probably showing an early diversification within dogs rather than separate cases of domestication. One lineage is from the Levant and is also found in African dogs; one is from the far northern area of Europe (Karelia) in Finland, Russia, Sweden, and the former Soviet Union; one lineage represents a Mesolithic population from the region of Lake Baikal in Siberia; one is from ancient America; and the fifth lineage includes the New Guinea singing dog and the Australian dingo, which are closer to unadmixed Southeast Asian dogs than anything else. These canine genomes were co-analyzed with seventeen sets of human genome-wide data that matched the age, geographic location, and cultural contexts of the ancient dogs. The researchers directly compared genetic relationships within the two species.[8]

The team found pairs in which evolutionary changes in dog and human populations mirrored each other and other cases in which they were decoupled. For example, an increase in the number of copies of the *AMY2B* gene, which facilitates starch digestion, is found in many dogs after the onset of agriculture, reflecting changes in the provisioning of dogs by humans. Humans, in parallel, underwent an increase in the number of copies of the *AMY1* gene, which also aids in digesting starch, though the increase of the *AMY1* gene is not as clearly linked to agriculture. The ancestral condition of only two copies of the *AMY2B* genes is retained in wolves, dingoes, and New Guinea singing dogs, including the fairly newly discovered highland population. However, another far northern dog, the Karelia bear dog—which is from roughly 9,600–8,500 years ago—already had four *AMY2B* genes, so perhaps it was adapted to eating starch before agriculture was widespread.[9]

FIFTEEN

Heading North

WHILE SOME PEOPLE were migrating southward from Beringia into the Americas, others continued farther north in Siberia with dogs that had very particular roles; these dogs were probably produced by deliberate breeding. These Arctic dogs were very special indeed; comparable specialization is not seen in early American dogs.

In addition to the hunter-gatherer-fisher lifeways manifested at the Yana Rhinoceros Horn site, another important part of the story of dogs and their remarkable adaptations to the far north took place in the Trans-Baikal region to the east and the Cis-Baikal region to the west of Lake Baikal, now in the Russian Federation. Despite quite a lot of archaeological work in the Trans-Baikal region, the reports often appear in local publications that are rarely accessible to English speakers. Robert Losey of the University of Alberta and collaborators have worked in this part of the world for many years and have begun summarizing the material and expanding access to the research for a broader audience. Their conclusions are remarkable.[1]

Cis-Baikal features two canid burials associated with large human cemeteries, such as the Shamanka II cemetery, which included ninety-six graves containing 154 people. Some graves also included canid remains, the oldest of which are close to 8,000 years old. A number of the dogs were of the approximate size of a Siberian husky or perhaps a chow chow: about 60 centimeters (cm) tall at the shoulder. Several dogs showed signs of healed injuries, suggestive of human care of injured dogs—care needed because of the strenuous use of dogs for transport of goods, from accidents, or from human punishments. These burials are among the earliest signs of canid involvement with humans in the region.

Though dogs were buried in graves also containing humans, no dogs were found buried there without humans. The Shamanka II dog was found in a grave with five humans, but they may have been later additions; there was a practice of reusing graves. The dog appears to have been the first body in the grave. It had suffered some rib fractures and trauma to its vertebrae in life, which may have been associated with a role as a beast of burden or with human-caused punishment.

Another similar cemetery, Lokomotiv, holds numerous burials of hunter-gatherer-fisher people. The mortuary treatment of canids in these cemeteries and others in the Cis-Baikal is highly variable, but some individual canids were buried with grave goods comparable to those interred with humans. These grave goods included objects of personal adornment, including jewelry made of animal teeth with holes bored in their roots so they could be suspended as necklaces or decorative elements, and household items, such as tools made of antler, stone, or bone. Sometimes needle cases bearing fine bone needles were included in graves. Bone needles were special items; they were essential for tailoring clothing from hides and were sometimes protected by needle cases made of hollow bird bones. Sometimes ochre was sprinkled on remains. Some graves held no grave goods, but others had up to 300 items. In general, graves might hold one, two, three, or four to eight human skeletons of men, women, and children. However, roughly one-quarter of the human skeletons were buried without the cranium, jaw, and last few upper vertebrae. The cranial portions appeared to have been intentionally removed before burial, but were not located.[2]

One Lokomotiv grave was an exceptionally rare find—it contained a large, aged male wolf, buried complete and alone in its own grave. What a striking individual that wolf must have been! The Lokomotiv wolf is particularly interesting because entire wolves were rarely buried. The skeleton, still in articulation, was buried in an oval pit with the head oriented to the south. The legs were slightly flexed and, very remarkably, between its ribcage and legs were the cranium, mandible, and first two vertebrae of an adult male human skeleton. No other isolated human cranium was found among the 124 individuals interred in this cemetery. The cranium appears to have been buried at the same time as the wolf; direct radiocarbon dates on some of the wolf bones yielded (statistically equivalent) ages of $7,320 \pm 70$ and $7,230 \pm 40$ years ago. The placement of the human cranium between the wolf's legs suggests the wolf may have been meant to protect the human in the afterlife, but there is no way to confirm or refute this speculation. Human and wolf may have died at the same time in a single event, but the cause of death is unknown. A human mandible (not from the same individual), a fibula, and a few scattered rib fragments and hand bones were found near the edge of the wolf's grave. It was common to reuse graves in Siberian cemeteries of this age, so these apparently random bones are interpreted as being from an earlier burial. Because graves were reused and remains from an earlier grave were often inadvertently mixed with those of later burials, it is not possible to determine whether any of the canid burials were actually single, isolated burials as defined by a typology developed by Angela Perri in her comparison of several large dog cemeteries. The dog burials do, in many respects, parallel the treatment of humans in the same cemetery.[3]

The Lokomotiv male wolf itself was formidable: very large, with a cranium 266 millimeters (mm) long, compared with the 216-mm-long cranium of a dog from the Shamanka cemetery. Its shoulder height was estimated to have been 74–79 cm, much bigger again than the Shamanka dog at 59–62 cm. The wolf was fully adult with markedly worn teeth, a few of which had been lost or split during life; the sockets had filled in with bone. An estimate of the individual's age at death was just over nine years, so it lived nearly twice as long as most wild wolves do. Genetically, the Lokomotiv wolf had a mitochondrial haplotype that was not matched by that of any previously published wolves' haplotypes, but it was closely similar to mitochondrial DNA (mtDNA) from Asian and Eurasian wolves. Analysis of the stable isotopes in the wolf's bones indicate that it fed mostly on deer, elk, and other wild ungulates, unlike the humans buried at Lokomotiv and the Shamanka dogs, which relied more heavily on fish and other aquatic foods. Thus, other than the burial with grave goods and a human cranium, there is no indication that the Lokomotiv wolf lived with or was intimately involved with humans. Why was this wolf buried? We do not know. But it is easy to imagine that such a

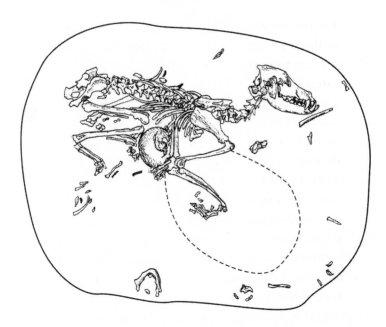

Lokomotiv is an important site in the Cis-Baikal region of Siberia that has been dated to more than 7,000 years before present. It includes a cemetery containing 124 humans, often in burials of multiple individuals or reused burials. The most unusual is the grave of a large, aged, intact male wolf, with ochre spread in an oval including the front right foreleg (*dotted area*). Between the animal's legs, close to its ribcage, are a human cranium, a jaw, and a few cervical vertebrae in articulation. A few scattered bones from previous burials are nearby but unrelated.

large and old wolf would have been known locally and admired for its hunting skill.

Generally, wolves were not so revered. There is one vaguely comparable grave at Khotorok, which included some bone fragments—but not a whole skeleton—that genetically matched a Russian wolf.

A number of complete adult canid crania and several partial or fragmentary crania were excavated in the Cis-Baikal cemeteries. Most suggest a body size similar to modern Siberian huskies, as do partial crania from Shamanka, but nowhere near as big as the Lokomotiv wolf. A group of fragments from one grave includes one dog cranium that is shorter in maximum length than the others, but it is a juvenile. It had sustained a puncture to its cranium and an injury to its pelvis, both of which had partially healed before death, evidence that this dog was cared for by people. These dogs grouped genetically with Clade I (or A) (see Chapter 6).

Ethnographies and cosmologies of northern Indigenous peoples emphasize that not only humans but powerful animals, landscapes, plants, and other beings may be considered ensouled. Proper, respectful treatment of an ensouled being at the time of death can ensure the recycling of souls back into new human beings; mistreatment could delay or disrupt such recycling. Bears figured most prominently among those whose remains were preserved, and crania were often used to represent the entire animal or species, according to Ivar Paulsen. This may take the form of burial in the ground (most common in Lapland), on platforms or in trees (usual in Siberia), or a third mode of burial in which the body was simply laid out on the ground, with or without a covering. Aquatic species were "buried" by being submerged in water. Burial of people and canids in pits in the ground was the usual treatment for the dead in the Baikal area from about 8,000 until about 6,800 years before present or slightly later, when formal burials of both people and dogs ceased for about 1,000 years. Human burials resumed about 5,000 years ago, but dog burials did not. About 3,400 years ago, pastoralists moved into the area with their sheep, goats, cattle, and horses. Were dogs not buried anymore because dogs were no longer useful in hunting? Or was it simply a cultural difference?

The Trans-Baikal cemetery on the other side of the lake tells a similar story with some interesting twists. The Trans-Baikal is a vast area; altogether it encompasses a region that is larger than any of the individual countries of Europe. Its southern border meets Mongolia and China, while the northern border is at the edge of the Patomskoe and North Baikal plateaus, a mountainous area covered in taiga-steppe. The western border is at the lake.

Several points are shown by the cemeteries in the Lake Baikal region and their contents. First, dogs commonly lived with people and were apparently considered worthy of elaborate burials in at least some instances. However, canids were not buried if humans were not buried. Second, in the Baikal region, humans relied both on hunting game and on aquatic resources; their remains showed bone composition similar to the dogs', suggesting the dogs were provisioned, while the Lokomotiv wolf had a distinctly different diet from local humans and ate more terrestrial prey and fewer aquatic species. Apparently wolves and dogs were no longer the same sort of animal. Third, the people who lived in the Cis- or Trans-Baikal were no longer the same sort of people they had been once. Either new people moved in at about 8,000 years before present and brought new mortuary rituals with them, or those rituals evolved, or the burial grounds were moved and have yet to be located. A few thousand years later, things changed again and neither people nor dogs were buried. Finally, at 3,400 years ago, human pastoralists moved into the area with new traditions of mortuary treatment. Seemingly dogs no longer interacted with humans on a daily, intimate basis, as they had when hunting and fishing provided most of the protein, and dogs were no longer buried as near-humans. What is seen in the Lake Baikal sites is that the treatment of dogs depended very closely on their interactions with humans and on the view those humans held about the afterlife.

The next remarkable archaeological site to provide stunning glimpses of people's views of the afterlife is in northeastern Siberia: Zhokhov Island, which is full of fascinating remains but offers few clear answers to many tantalizing questions. Here, once again, dogs seemed to play a crucial role in human survival; the Zhokhov site is one of the northernmost known sites with evidence of human habitation and an unusual subsistence strategy. Now a small island, Zhokhov was once part of a large coastal lowland covering the area of the New Siberian Islands today; the period of interest preserved on Zhokhov is about 9,500–8,000 calibrated years before present. Vladimir Pitulko and a team from the Russian Academy of Sciences in St. Petersburg began excavating at Zhokhov in 1998–1999 and resumed between 2000 and 2012 as an interdisciplinary team working with Edmund Carpenter, an American specialist in Siberian cultures, and scientists from the Smithsonian Institution.[4]

Almost 25,000 fossil specimens (dominated by remains of reindeer and polar bear) and 19,000 lithic specimens have been recovered from the site. Many of the tools were made of obsidian from a locality 1,500 kilometers away. This implies a long-distance trade network that must have been difficult to maintain in the high Arctic. More than 300 objects were made of antler, mammoth ivory, and bone; approximately 1,000 objects were made of wood, and there were a few woven and birch bark artifacts. Extreme cold preserves artifacts well, even those that usually decay.

From this wealth of evidence, Pitulko and the team deduced that the Zhokhov people hunted hibernating polar bears in winter by finding their dens, which were basically snow caves.

Pitulko's reconstruction of how the polar bear was hunted is extraordinary. The dogs helped the people find the dens by scent; some were maternal dens and some simply hibernation dens for the bears. People closed off the entrances with snow and used the dogs to frighten the adult polar bears into going deep into the dens and breaching through the roofs. If the snow was deep enough, the bears could not immediately climb out through the roofs and the barking dogs encouraged the bears to stick their heads out through a small hole to check for danger. Waiting hunters could kill the bears from a distance with spears or arrows to the head, leaving distinctive damage on the crania. If there were any juvenile bears, they were trapped in the den and could be dispatched with less risk. A key aspect of this hunting strategy is that the powerful and agile bear was more or less confined by the size and shape of the den because a free-moving polar bear is very large indeed and likely to attack a hunter or his dogs.

Because reindeer were migratory and relatively rare during winter, polar bears were a critically important fallback food. However, nobody would claim that hunting polar bears was an easy occupation. In spring, the Zhokhov people preferentially hunted reindeer, particularly favoring hunting offspring, who were large enough at one year to be the size of adults but had more tender and more flavorful meat. However, most reindeer were killed in the fall, when they would be relatively fat in preparation for winter. Calculating how many individual animals are represented in the collections shows that roughly twice as many reindeer (245) were killed as polar bears (130). Bears yielded three times as much meat per individual as reindeer, but they were also far more dangerous. Apparently reindeer were tastier, and they also yielded antlers, which were often used in the manufacturing of various tools. With both types of prey, butchery was carried out at the kill sites and only those bones bearing preferred cuts of meat or rich marrow were brought back to the camp.

Canids that Pitulko and colleagues identify confidently as dogs constitute only about 0.5 percent of the identifiable fauna remains (151 specimens), but they played a key role in the hunting and subsistence strategy that enabled people to live in the high Arctic for much of the year. Without the ability to successfully hunt polar bears in winter, subsistence in the region might well have been impossible. And without large dogs, hunting polar bears would have been an extremely hazardous occupation.

The remains of a dog the size of modern huskies or malamutes and a few items that were parts of sophisticated sleds (such as part of a runner, stanchions, and toggles from the harness) were identified from the site. Before excavations at Zhokhov, dog sleds were seen as an iconic part of the Inuit culture, which started only about 1,000 years ago. The finds at Zhokhov—well preserved by permafrost—contradict this belief, showing that sophisticated dog sled transport was used 7,000 years earlier than had been thought. And dog sleds were probably essential to the long-distance trade network the Zhokhov people maintained.

The dogs at Zhokhov came in two distinct sizes, shown clearly in the two largely intact canid skulls from the site. Other remains include jaws, a few vertebrae, and various limb bones of probably thirteen individual dogs. Analysis of the control region of mtDNA of the fossil specimens showed the Zhokhov dogs carried haplotypes well known from Eurasian dogs in Clade A (or I); the canids from Zhokhov are not wolves. Pitulko transformed various measurements of the crania of the Zhokhov canids into indices capturing the proportional differences between wolves and dogs, using samples of thirty-two sled dogs collected a hundred years ago in eastern Siberia and twenty-four Siberian wolves as references. One of the two Zhokhov crania unquestionably groups with sled dogs, whereas the other was closer to, but not within, the cluster of wolves.

The size and body weight of sled dogs were calculated from other bones or teeth of the Zhokhov remains. The larger cranium was estimated to have come from a dog with a body weight of 30 kilograms (kg). This is reaching the upper limit of size at which a sled dog can have the needed strength and still be able to dissipate heat from running and pulling loads effectively. Ten of the Zhokhov canids have estimated body weights between 16 kg and 25 kg, which fits the modern breed standard for sled dogs. The larger dog at Zhokhov is the size of modern malamutes or Greenland sled dogs; the smaller ones match Siberian huskies in size. Pitulko hypothesizes that the bigger dogs were used for hunting and finding polar bear dens, whereas the smaller dogs were used to pull sleds loaded with the parts of the prey that were carried from the kill sites to the camp site.

Finally, Pitulko and his team note that dog skulls were carefully removed from the rest of the body, leaving only light cutmarks, without the predictable and consistent damage caused by removing and consuming soft tissues, such as the tongue, brain, and chewing muscles, from reindeer or bear skulls. They suggest that the difference between this activity and the processing of polar bears and reindeer may have been related to some sort of ritual or belief concerning dogs, but what that was is not clear. What is clear is that the well-preserved collection of remains from Zhokhov has transformed our knowledge of, and evidence for, the deliberate breeding of dogs for specific jobs almost 8,000 years ago. These jobs enabled people who worked with dogs to adapt to and live in the high Arctic, hunting large and ferocious bears as well as reindeer. The amazing preservation of specimens at Zhokhov has not only enabled this surprising and detailed reconstruction to be made but has also revealed the existence of a sustainable economic strategy in an unforgiving habitat. However, additional adaptations to an Arctic way of life would not develop in these dogs until later.

In general, humans first inhabited northern Alaska, Canada, and Greenland starting around 6,000 years ago, but they were not all members of the same cultural group. The history of people and dogs in the American Arctic is complex. The earliest culture is known as the Early Paleo-Eskimos (Pre-Dorset / Saqqaq), followed by the Late Paleo-Eskimos (Early Dorset, Middle Dorset, and Late Dorset) and the Thule cultures. The Pre-Dorset people occupied the eastern Canadian Arctic from about 3200 to 850 BCE (calibrated). The subsequent Dorset culture is divided into early (500–1 BCE), middle (1–500 CE), late (500–1000 CE), and terminal phases (from 1000 CE onward). The Thule were ancestors of the Inuit people who spread rapidly eastward with their dogs from Siberia to Greenland starting about 1000 CE, a feat attributed largely to their invention of the umiak and kayak for moving across water and superior sled technology for transportation on ice or snow. This expansion may have spread the genes of the powerful Zhokhov dog to the east and into Greenland sled dogs, which still carry them.[5]

Modern Greenland sled dogs and the Zhokhov dog share a number of genetic adaptations useful to sled dogs. Not surprisingly, this group carries very few copies of the $AMY2B$ gene that is carried in multiple copies by most dogs and aids in the digestion of starch. Because Arctic peoples historically have had little access to starchy foods, this is to be expected. However, the MGAM haplotype that is found in high frequency in wolves, but not in most dogs, is also uncommon in this sample. Other genomic resemblances between the Zhokhov dog and Greenland sled dogs are adaptations to cold and to maintaining an appropriate intake of oxygen for muscle contraction during prolonged exercise. However, the modern sled dogs also show adaptations to a high intake of fatty acids and to clearing cholesterol from the blood, an adaptation suggestive of being provisioned on blubber that is not seen in the Zhokhov dog. Similar adaptations to diets high in fatty acids have been reported among the Inuit and other human populations that moved into the eastern Arctic, among whom modern Greenland sled dogs have lived for the past thousand years or so.

Much earlier than the Inuit expansion across the Northern Hemisphere, about 20,000 years ago, Beringian gray wolves had undergone a parallel eastward expansion across the Northern Hemisphere. However, the Beringian gray wolves contributed little genetically to Greenland sled dogs, which, more than other groups of sled dogs, have been kept from cross-breeding with wolves or different sled dogs. Although ethnographic anecdotes of wolves and sled dogs cross-breeding suggest that such events were common, the hybrids were more difficult to handle in the domestic context and performed more poorly in pulling sleds. Current evidence does suggest that Beringian wolves were the ancestors of most or all of the Northern Hemisphere gray wolves that gave rise to domesticated dogs.[6]

After living in Zhokov, did these innovative Arctic hunters and their deliberately bred dogs move into Beringia and down into the Americas? Mikkel-Holger Sinding and a large team of colleagues extracted a nuclear genome from a Zhokov dog and compared it with that of a wolf from the Yana Rhinoceros Horn site and with genomes from ten modern sled dogs from Greenland to answer this question.[7] The Zhokhov dog was closely related to the Greenland sled dogs and shared the low number of copies of starch-digestion genes. However, the Greenland dogs apparently acquired genes involved in eating a high-fat diet that are similar to genes reported among Inuit peoples after the Greenland dogs diverged from the Zhokhov dog, which does not carry such genes. The issue is why we do not see the dogs that the first group of Siberians moving into Beringia and down into the Americas should have brought with them. Clearly, working closely with dogs bred for whatever job they performed was a tradition in the high Arctic of Siberia well before humans first arrived in the Americas, but how exactly those people and their dogs migrated into the Americas remains unclear. People could have moved south through the ice-free corridor or along the kelp highway in boats starting about 16,000 or 15,000 years ago.

A team led by Elizabeth P. Murchison, Greger Larson, and Laurent A. F. Frantz generated complete mitochondrial genomes from seventy-one ancient American or Siberian dogs from the past 9,000 years; they also considered 145 dogs with mtDNA that had been reported in the literature. The dogs formed a phylogenetic tree within the most common dog clade (Clade A or I). All dogs in the sample shared an ancestor related to the Zhokhov dogs at about 15,000 years ago. This clearly showed that dogs were in the Americas before European contact. However, they were not derived from North American wolves but probably from Siberian wolves.[8]

Their sample also included the oldest directly dated archaeological dogs in the Americas from the Koster and Stillwell sites in Illinois. The Koster dogs and the Stillwell dog were recently redated to about 10,000 years ago and appear to have been intentionally buried. The Koster dog appears to be more gracile and less robust than the Stillwell dog. The dating implies either that hunter-gatherers from the Arctic migrated down into the Americas without their dogs, which seems an odd choice, or that the dogs were not in the Americas in large numbers until 5,000–6,000 years after the first migration, perhaps accompanying a second wave of humans. Unfortunately, the Stillwell dog's remains did not yield any DNA. The dogs from the two American sites are estimated to have been Zhokhov or Baikal-region dogs: between 439 and 517 cm tall at the shoulder and weighing only about 17 kg. This makes them much lighter in weight than the large dog at Zhokhov is estimated to have been, but close in height and not far from the American Kennel Club standard for Greenland sled dogs, like the smaller Zhokhov dogs. But being a sled dog was apparently not a job of much value in Illinois 10,000 years ago. Being a companion might have been.

This study of American dogs before European contact confirmed the results of a previous study by Jennifer Leonard in 2002. She and her team showed that American dogs were not independently descended from North American wolves, based on the analysis of mtDNA segments of thirty-seven native dogs from archaeological sites in Mexico, Peru, and Colombia. Her sample included eleven pre–European contact dogs from Alaska. A few sequences from the Eskimo dog, the Mexican hairless, the Alaskan husky, the Newfoundland, and the Chesapeake Bay retriever, plus the Australian dingo and the New

Guinea singing dog from Oceania, were added to the sample.
All grouped with Eurasian dogs. Leonard's colleagues' analysis
suggested that at least five lineages of dogs came south with
humans from the far north. The subsequent near-disappearance
of most of these lineages shows how strongly a colonial culture can suppress Indigenous cultures—and their dogs.[9]

A recent paper has argued that the genetic and geographic
data on Siberian dogs so closely mirrors the movements of
people as they migrated into the Americas that surely Siberia
was the beginning of dog domestication. A group of prominent researchers on dog genetics and human genetics collabo-
rated on this work. Unfortunately, there are no radiometrically
dated sites where the earliest clearly domesticated dogs can be
found. Rather than rely on the radiometric dating of archaeological or paleontological finds, the team dates their scenario
only on the basis of observed mutations, which tends to yield
variable results. Thus, the team concludes:

> Here, by comparing population genetic results of
> humans and dogs from Siberia, Beringia, and North
> America, we show that there is a close correlation in
> the movement and divergences of their respective lin-
> eages. . . . It suggests that dogs were domesticated in
> Siberia by ~23,000 y[ears] ago, possibly while both
> people and wolves were isolated during the harsh
> climate of the Last Glacial Maximum. Dogs then ac-
> companied the first people into the Americas and trav-
> eled with them as humans rapidly dispersed into the
> continent beginning ~15,000 years ago.

But there are no sites in their analysis dated to 23,000 years
ago. The Yana Rhinoceros Horn site in Siberia is older (33,000

years), the Lake Baikal sites are much more recent (8,000–9,500 years old), and the Zhokhov polar bear–hunting dogs are about 9,500 years old. The earliest dog remains in North America are only 10,000 years old. Quite possibly the notion of domesticating dogs and working with them traveled southward with Siberians to populate the Americas. The team's conclusions are stunning and may well be generally correct, but they do not support any hypothesis about how dogs or semidomesticated canids got from Siberia or the Americas to Southeast Asia or Greater Australia.[10]

SIXTEEN

To the End of the Earth

As humans moved south in the Americas, conditions varied and so did canids. In an interesting parallel to Australia, there were no canids at all in South America until 3–2.5 million years ago, when terrestrial species could first walk to South America via the upraised Isthmus of Panama. However, South America is now home to some endemic and rather fox-like canids—often known as *zorros*—that were fundamentally isolated from the time when they first entered the continent until European colonization. As a group, these species are unlike the canid fauna of other geographic regions in terms of the details of their anatomy. Early colonial accounts of the endemic canids of the Amazon Basin and the Caribbean mentioned repeatedly that these species do not bark. Many authors mentioned two types: one was probably derived from Spanish mastiffs brought in by colonial settlers and the other was a small white dog of unknown ancestry. They were used to hunt the large endemic rodent known as the *hutia.* The dogs were also considered good eating, like young goats.[1]

South American endemic "dogs" are not very well known outside the continent and are not actually dogs; they have never been domesticated. They include the very long-legged maned wolf (*Chrysocyon brachyurus*), a striking animal with a reddish coat, long black legs and muzzle, and a fluffy, fox-like tail. It is largely solitary and feeds on insects, small rodents, and birds in grassland areas. The endangered and rare small-eared dog (*Atelocynus microtis*) is a slender species adapted to the tropical rainforest that avoids people and disturbed habitats. The bush dog (*Speothos venaticus*) is a short-legged, stocky animal that is primarily a pack hunter of agouti and paca. It frequents riverine habitats and is considered semi-aquatic, in part because it has webbed toes. Another one of the endemic species of South America is the crab-eating fox (*Cerdocyon thous*), a short-legged, ground-dwelling omnivore that lives in forested areas, eating small mammals, insects, fruits, crabs, and frogs.[2]

South America also boasts a group of six species known as foxes or false foxes, in the genus *Lycalopex* or *Pseudalopex,* though some scholars group them with the Falkland Islands wolf in *Dusicyon. Dusicyon australis,* the largest native South American canid, is now regrettably extinct; it was about the size of a coyote (about one meter high at the shoulder). Darwin saw and interacted with this species on his famous voyage on the HMS *Beagle* and found it remarkably tame, wondering if it had actually been domesticated. It is known only from the Falklands.

There are numerous anecdotal historical reports of the taming of South American foxes and the breeding of hybrids with domestic dogs; these have been believed, or discredited, by various

The indigenous South American canids are an unusual group of canids not closely related to foxes, wolves, or dogs. *Opposite, above:* The crab-eating fox (*Cerdocyon thous*) often lives near small bodies of water in central South America. *Opposite, below:* The bush dog (*Speothos venaticus*) is a rare animal with a wide range. *Above:* The maned wolf (*Chyrsocyon brachyurus*) is a beautiful animal that looks like a fox on stilts, though it is not closely related to foxes or wolves. Its long legs appear to be an adaptation to hunting in tall grasses.

authors. Contradictory statements call into question the veracity of these reports. Some label the endemic canids as mute or unable to bark, while others offer descriptions of different vocalizations (growls, yelps, howls). One writer reports they are easy to tame, but another says they are "rarely seen because they flee from humans." There is no convincing evidence of the domestication of any of the South American canids. And although bush dogs were originally collected from the limestone caves of Brazil, where some human remains were buried, there is no hint that the bush dogs were domesticated or deliberately buried.

Charles Darwin was personally puzzled by the fact that the only terrestrial species in the Falklands was a wolf. The islands are about 480 kilometers from the nearest land mass, which may have informed his speculation that they might have been brought to the Falklands by people. The closest relative to *Dusicyon australis* was a small, fox-like, extinct canid (*Dusicyon avus*) that lived in Argentina. Modern research into the genetics of *Dusicyon* confirmed that the two were indeed closely related, having separated only about 16,000 years ago. The finding that the distance between the Falklands and Argentina had shrunk to only about 20 kilometers at that time suggested that the wolf could have crossed to the island during an icy freeze without human assistance. It seems that fox-like canids migrated into South America and then rapidly diversified into various niches based on diet and habitat. A few fossils of the North American dire wolf (*Canis dirus*) and the gray fox (*Urocyon cinereoargentus*) have also been found in South America. Other wolves did not make it into the Southern Hemisphere and thus could not evolve into dogs.[3]

What this means is that true domestic dogs almost certainly did not enter South America in numbers until European colonizers brought them. Siberians and other early immigrants into the Americas did not, apparently, often bring dogs with them into the Americas, not even those who had specially bred dogs when they lived farther north. Prehistoric human burial grounds are known from South America and have yielded many skeletal remains, such as the famous Luzia cranium from Lagoa Santo, dated to more than 9,000 years ago.

Geneticists have attacked the problem of South American settlement by extracting whole genome DNA from forty-nine ancient South Americans from Brazil, Belize, Peru, Argentina, and Chile. They established a genetic link between these early

South Americans and the Anzick Boy from Montana. That child was found with Clovis artifacts and dated to between 12,900 and 12,700 years before present; he was related to many of the earliest South American samples. But the more recent samples from South America contained DNA links to the second branch of humans thought to have given rise to all native Americans from North America, suggesting that the early populations of South America were replaced by others starting about 9,000 years ago. Still other recent analyses suggest that some ancient South Americans came from a still-unidentified population with a greater representation of genes from Australia or Asia, which may account for the unusual appearance of some crania, such as the Luzia fossil. The isolation of South America apparently affected the evolution of both canids and humans as isolation in Greater Australia affected both Aboriginal peoples' and dingoes' evolution.[4]

What happened when true domesticated dogs entered South America? Any attempt to trace the movement of dogs and people in South America must deal with the striking differences between the use of dogs among living hunter-gatherers and the use of dogs in the colonial past. During the conquests by Spanish and other European invaders, dogs were instruments of conquest. Dogs, particularly aggressive Spanish mastiffs, were trained and used to chase and kill Indigenous people to the point of eliminating entire villages or Indigenous groups. The dogs actually fed upon the bodies of those they killed. They were effective weapons that were repeatedly and deliberately applied against Indigenous people. Dogs were one part of a conquest technology package that included horses, armor, firearms, and metal swords. European dogs were initially perceived as terrifying by many Indigenous people because they had never before seen such animals, even if there were native canid species in the area (such as in the Brazilian Amazon). Paradoxically, European dogs were a much-requested gift for use as a hunting accessory by Indigenous peoples when they had contact with Europeans. In a stunning parallel to the reception of true dogs in Greater Australia, dogs were adopted rapidly and eagerly by the Indigenous peoples of South America. Dogs were also known to colonizers for their ability to forge emotional links between opposing human groups. As recently as 2013, Felipe Vander Velden was told the story of Cândido Mariano da Silva Rondon, who had explored the region during the first rubber boom (1906–1909). On one occasion, Rondon's much-loved dog wandered off and was found by locals of the Puruborá people, who recognized that he was a "white man's dog" and escorted the dog back to his owner. Both Rondon and the dog were delighted to be reunited, and henceforth Rondon "tamed" the Puruborá and liked them. I cannot help but wonder what he meant by "tamed."[5]

Vander Velden recounts a somewhat similar story of the first European dog known in the country of the Karitiana people, who were drawn into the rubber industry that infiltrated their territory. The first dog was brought to their village by a traveling merchant who marketed manufactured goods as he wandered through the area buying forest products. (Another version of the story identifies the dog owner as an Indigenous leader.) The small white dog stayed in the village and proved to be a good hunter, making the Karitiana anxious to obtain more dogs so they could breed more hunting assistants. It is said that everyone in the village liked the dog and was not afraid of it. It became known as a "domestic jaguar" for its ferocity in hunting, while it was still recognized as a member of the human family, like a son or daughter.[6]

The primary point Vander Velden is making is that dogs played a critical role in easing tensions between different peoples, as social lubricants in tense interactions, and in facilitating communication among strangers. He even suggests that the native "dogs"—in the form of the indigenous foxes *Cerdocyon* (the crab-eating fox) and various species of *Lycalopex*—may have paved the way for the acceptance of a true domestic dog in remote regions of South America when they were brought in by Europeans. Despite the memories of European dogs having been used as instruments of conquest, they were welcome. Vander Velden identifies two reasons. The first is the hunting skill of the dogs, which was much admired; the second is their tendency to form close emotional bonds with humans.

In a quantitative study, anthropologist Jeremy Koster found that hunting with dogs in forested regions of Nicaragua increased the success rates of Indigenous hunter-gatherers about as much as having a firearm did. Koster's data permitted a statistical analysis of the costs and benefits of hunting with dogs, including the dogs' need for meat and their appallingly high mortality rate. Koster found that 49 percent of adult dogs at his study site died annually and almost half of all puppies died as neonates. He established that dogs who were considered better hunters also received more food and other care. Of course they did.[7]

Koster and Vander Velden both write of the friendliness of dogs, of their ability to work with humans and form deep emotional bonds. This is the same sort of emotional connection that Clive Wynne writes about in *Dog Is Love* and that Brigitte vonHoldt is researching in her comparisons of the behavioral effects of certain genetic mutations in dogs that seem to cause friendliness and extreme openness to strangers or novelties. Similar emotional ties are implied in the extraordinary

What makes a dog a dog? The relationship between the dog and people.

lifestyle of the far northern Arctic dogs at Zhokhov, which were intentionally bred and buried. The friendliness that makes dogs appealing emotionally is combined with varied physical traits that enable dogs to play very different but essential roles in human survival, whether in hunting polar bears, tracking the nine-banded armadillo, smelling out underground water, pulling sleds, or warning of ghosts and evil spirits. It is this surprising combination of behavioral and physical traits that has enabled dogs to become valued companions in many different circumstances—that has allowed humans to breed for size, strength, furriness, or color, and to develop special means of communication suitable to different hunting strategies.[8]

Postscript

What seems to be true, in reviewing the migrations and expansions of people and dogs around the world, is that their movements were strongly affected by biogeographic factors. Where humans entered a continent that could not be reached without maritime skills, doglike animals could not enter either without those skills. Canids did not acquire those skills directly, but rather by forming intimate emotional ties to humans. Thus, when humans migrated into new territories or ecosystems, they often brought their canine companions with them.

In the case of Australia, where humans faced exceptionally arid conditions and a limited water supply, humans entered the continent first and developed cultural mechanisms for survival. The clever adaptations that enabled humans to learn enough about the habitat to survive were firmly in place by the time dingoes arrived. But what dingoes did not do, in their own adaptations to the harshness of living in Australia, was evolve mechanisms or skills that helped the First Australians to survive. Dingoes were not essential to the lives of Indigenous Australians, nor were the Indigenous Australians vital to dingo

life, though the anthropogenic habitat offered important resources. The deep need and urgent pressure to develop strong
emotional bonds and crucial communication skills simply did
not apply to dingoes and humans in Australia.

Similarly, the first people to enter the Americas—and particularly the tropical forests in much of South America—did
not bring their dogs with them, and only much, much later saw
the advantages that working collaboratively with canids might
impart. But neither in South America nor Australia did the Indigenous people domesticate the canids available to them,
though both were impressed by the domesticated dogs that colonial settlers had.

My assessment is that neither dingoes nor the endemic canids of South America were domesticated dogs and they were
not subsequently domesticated. They show us that the essential elements of dogginess are separable and highly varied, and
that this underlies the great success dogs have had in the world.
But the First Australians never interacted with dingoes in ways
that favored their domestication into dogs, and dingoes never
acted toward humans in ways that led to their wholehearted
adoption as members of the family. Dingoes were never fully
incorporated into traditional Indigenous culture, although they
assumed key roles in Dreamtime stories. The bond between
Indigenous Australians and dingoes was expected to be a temporary alliance, not a permanent one. The Indigenous peoples
of South America do not seem to have domesticated their local
canids, either, and did not have native wolves of any kind.

I have seen clearly in my research that people who like, ad-
mire, and understand dingoes *do* develop extraordinary, long-
lasting bonds with them. Most people lack the skill or will to do
this and to integrate them completely into their lives. When they
first met with the people of Australia, dingoes were partway

to becoming dogs but never got there. We find it difficult to trace where dingoes came from, other than somewhere in Asia, and thus we find it unclear what exactly they are specialized to do, other than being dingoes.

Why were they brought along on boats heading to the next island or Greater Australia? We don't really know. As long as dingoes fill an ill-defined, liminal niche—sometimes food, sometimes companion; occasional protector, occasional predator; sometimes child substitute, sometimes mischief maker or thief; playmate or occasional nuisance, occasional hunting companion, occasional target—we will have trouble answering this question.

The same might be said of the endemic canids of South America. Their unusual qualities parallel the oddness of some of the ancient humans that have been found there. I suspect there was some sort of biogeographic barrier at work, but I don't know what it was.

The story of each migration or radical expansion of territory that humans undertook posed a different set of challenges each time. Where it was colder, but animals and plants were still vaguely familiar, humans and their dogs were forced to adapt to those hostile conditions and to differences in the behavior of the plants and animals upon which they relied. In all probability, it wasn't a conscious decision, but a simple do or die. In open grasslands or tundra, moving fast and hunting in a coordinated fashion was essential. A different kind of strategy worked better in dense, humid, tropical forests. When people and dogs moved into habitats that were extraordinarily cold and dangerous, we developed truly amazing adaptations to survive together and obtain and digest the available foods. The depth of the bond that grew between people and dogs became even more obvious. Religious and spiritual beliefs flourished about the connection between the animals we hunted and the animals we lived with; sometimes we were one community. Burying and honoring animals according to their roles in our lives assumed a much greater importance than it had previously. We humans began to sacrifice a good deal and go to a lot of trouble to ensure that we behaved appropriately toward the animals that nourished and protected us. We developed a sort of thoughtful morality that governed our behaviors and our expectations of our companions.

This book has been a fascinating journey for me, a puzzle and a pursuit of understanding that has sometimes led to intellectual struggles and frustration. I may be mistaken in the threads of arguments or in my understanding, but I am grateful for the effort that this topic has required of me. More clearly than ever, I think I see how it is that we live with other animals and how mutually dependent we are. Humans are part of the same ecosystems as canids; that is why humans often bring along the dogs and other animals with which they have negotiated a deeply felt alliance. *Dogs are people,* in the sense that they have individual ideas, perceptions, emotions, and opinions. They have emotion-based moral standards for acceptable behavior and some, maybe all, have a sense of fairness. It should be said that *people are animals, too,* in that we are moved by physical and psychological needs not too dissimilar from those we recognize clearly in other species. By choosing to live with us and to communicate with us, dogs have given us a tremendous opportunity for learning how to survive and thrive together.

NOTES

ACKNOWLEDGMENTS

ILLUSTRATION CREDITS

INDEX

Notes

Preface

1. C. Hall, "Why Zebra Refused to Be Saddled with Domesticity," TheConversation.com (https://theconversation.com/why-zebra-refused -to-be-saddled-with-domesticity-65018, September 14, 2016).
2. R. Coppinger and L. Coppinger, *Dogs: A Startling New Understanding of Canine Origin, Behavior & Evolution* (New York: Scribner, 2001).

1. Before Dogs

1. Bronwen Dickey, *Pit Bull: The Battle over an American Icon* (New York: Knopf, 2016).
2. E. Matisoo-Smith, "The Human Colonisation of Polynesia. A Novel Approach: Genetic Analyses of the Polynesian Rat (*Rattus exulans*)," *Journal of the Polynesian Society* 103, no. 1 (1994): 75–87.
3. R. Wayne, "Molecular Evolution of the Dog Family," *Trends in Genetics* 9, no. 6 (1993): 220.
4. K.-P. Koepfli, J. Pollinger, R. Godinho, et al., "Genome-wide Evidence Reveals That African and Eurasian Golden Jackals Are Distinct Species," *Current Biology* 25 (2015): 1–8.
5. C. Daujeard, G. Abrams, M. Germonpré, et al., "Neanderthal and Animal Karstic Occupations from Southern Belgium and South-eastern France: Regional or Common Features?" *Quaternary International* 411, part A (2016): 179–197.

6. K. Lohse and L. Franz, "Neandertal Admixture in Eurasia Confirmed by Maximum-Likelihood Analysis of Three Genomes," *Genetics* 196 (2014): 1241–1251.

7. V. Slon, F. Mattazoni, B. Vernot, et al., "The Genome of the Offspring of a Neanderthal Mother and a Denisovan Father," *Nature* 561 (2018): 113–116; L. Chen, A. Wolf, W. Fu, et al., "Identifying and Interpreting Apparent Neanderthal Ancestry in African Individuals," *Cell* 180, no. 4 (2020): 677–687.

8. T. Higham, K. Douka, R. Wood, et al., "The Timing and Spatiotemporal Patterning of Neanderthal Disappearance," *Nature* 512 (2014): 306–309.

9. M. Germonpré, M. V. Sablin, R. E. Stevens, et al., "Fossil Dogs and Wolves from Palaeolithic Sites in Belgium, the Ukraine and Russia: Osteometry, Ancient DNA and Stable Isotopes," *Journal of Archaeological Science* 36, no. 2 (2009): 473–490; M. Germonpré, M. Lázničková-Galetová, and M. Sablin, "Palaeolithic Dog Skulls at the Gravettian Předmostí Site, the Czech Republic," *Journal of Archaeological Science* 39, no. 1 (2012): 184–202.

10. O. Thalmann, B. Shapiro, P. Cui, et al., "Complete Mitochondrial Genomes of Ancient Canids Suggest a European Origin of Domestic Dogs," *Science* 342 (2013): 871.

11. H. Bocherens, D. Drucker, M. Germonpré, et al., "Reconstruction of the Gravettian Food-web at Předmostí I Using Multi-isotopic Tracking (^{13}C, ^{15}N, ^{34}S) of Bone Collagen," *Quaternary International* 359–360 (2015): 211–228.

12. K. Prassack, J. DuBois, M. Lázncková-Galetová, et al., "Dental Microwear as a Behavioral Proxy for Distinguishing between Canids at the Upper Paleolithic (Gravettian) Site of Předmostí, Czech Republic," *Journal of Archaeological Science* 115 (2020): 105092–105102.

13. A. R. Perri, "Hunting Dogs as Environmental Adaptations in Jōmon Japan," *Antiquity* 90, no. 353 (2016): 1166–1180.

14. P. Shipman, *The Invaders: How Humans and Their Dogs Drove Neanderthals to Extinction* (Cambridge, MA: Belknap Press of Harvard University Press, 2015).

2. Why a Dog? And Why a Human?

1. M. Derr, *How the Dog Became the Dog: From Wolves to Our Best Friends* (New York: Overlook Duckworth, 2011), 20.

2. C. Mason, personal communication to author, 1976.

3. A. Miklósi, E. Kubinyi, J. Topál, et al., "A Simple Reason for a Big Difference: Wolves Do Not Look Back at Humans But Dogs Do," *Current Biology* 13 (2003): 763–766; K. Lord, "A Comparison of the Sensory Development of Wolves (*Canis lupus lupus*) and Dogs (*Canis lupus familiaris*)," *Ethology* 119, no. 2 (2013): 110–120.

4. C. Wynne, *Dog Is Love: Why and How Your Dog Loves You* (New York: Houghton Mifflin Harcourt, 2019).

5. R. Coppinger and L. Coppinger, *Dogs: A Startling New Understanding of Canine Origin, Behavior and Evolution* (New York: Simon and Schuster, 2001).

3. What Is Dogginess?

1. C. Wynne, *Dog Is Love: Why and How Your Dog Loves You* (New York: Houghton Mifflin Harcourt, 2019).

2. B. VonHoldt, J. Pollinger, K. Lohmuelle, et al., "Genome-Wide SNP and Haplotype Analyses Reveal a Rich History Underlying Dog Domestication," *Nature* 464 (2010): 898–903, 901.

3. B. Hare, M. Brown, C. Williamson, and M. Tomasello, "The Domestication of Social Cognition in Dogs," *Science* 298, no. 5598 (2002): 1634–1636.

4. One Place or Two?

1. J. Krause, Q. Fu, J. Good, et al., "The Complete Mitochondrial DNA Genome of an Unknown Hominin from Southern Siberia," *Nature* 464 (2010): 894–897; D. Reich, R. Green, and S. Pääbo, "Genetic History of an Archaic Hominin Group from Denisova Cave in Siberia," *Nature* 468 (2010): 1053–1060.

2. Reich et al., "Genetic History of an Archaic Hominin Group."

3. M. Caldararo, "Denisovans, Melanesians, Europeans and Neandertals: The Confusion of DNA Assumptions and the Biological Species Concept," *Journal of Molecular Evolution* 83 (2016): 78–87; J. Hawks and M. Wolpoff, "The Accretion Model of Neandertal Evolution," *Evolution* 55, no. 7 (2001): 1474–1485; J. Hawks and M. Wolpoff, "Brief Communication: Paleoanthropology and the Population Genetics of Ancient Genes," *American Journal of Physical Anthropology* 114 (2001): 269–272.

4. S. Brown, T. Higham, V. Sion, et al., "Identification of a New Hominin Bone from Denisova Cave, Siberia, Using Collagen Fingerprinting and Mitochondrial DNA Analysis," *Scientific Reports* 6 (2016): 23559.

5. F. Chen, F. Welker, C-C. Shen, et al., "A Late Middle Pleistocene Denisovan Mandible from the Tibetan Plateau," *Nature* 569 (2019): 409–412; S. Bailey, J-J. Hublin, and S. Antón, "Rare Dental Trait Provides Morphological Evidence of Archaic Introgression in Asian Fossil Record," *Proceedings of the National Academy of Sciences* 116, no. 30 (2019): 14806–14807.

6. G. Scott, J. Irish, and M. Martinón-Torres, "A More Comprehensive View of the Denisovan 3-Rooted Lower Second Molar from Xiahe," *Proceedings of the National Academy of Sciences* 117, no. 1 (2019): 37–38.

7. Reich et al., "Genetic History of an Archaic Hominin Group."

8. P. Qin and M. Stoneking, "Denisovan Ancestry in East Eurasian and Native American Populations," *Molecular Biology and Evolution* 32, no. 10 (2015): 2665–2674.

9. D. Rhode, D. Madsen, J. Brantingham, and T. Dargye, "Yaks, Yak Dung, and Prehistoric Human Habitation of the Tibetan Plateau," in D. B. Madsen, F. H. Chen, and X. Gao, eds., *Late Quaternary Climate Change and Human Adaptation in Arid China* (Amsterdam: Elsevier, 2007), 205–224; M. Hanaoka, Y. Droma, B. Bassnyat, et al., "Genetic Variants in EPAS1 Contribute to Adaptation to High-Altitude Hypoxia in Sherpas," *PLoS ONE* 7, no. 12 (2012): 50566.

10. R. M. Durbin, G. R. Abecasis, R. M. Altshuler, et al., "A Map of Human Genome Variation from Population-Scale Sequencing," *Nature* 467 (2010): 1061–1073.

11. Durbin et al., "A Map of Human Genome Variation."

12. S. Kealy, J. Louys, and S. O'Connor. "Least Cost Pathway Models Indicate Northern Human Dispersal from Sunda to Sahul," *Journal of Human Evolution* 125 (2018): 59–70.

5. What Is Domestication?

1. B. Hesse, "Carnivorous Pastoralism: Part of the Origins of Domestication or a Secondary Product Revolution?" in R. Jameson, S. Abouyi, and N. Mirau, eds., *Culture and Environment: A Fragile Coexistence* (Calgary: Proceedings of the 24th Annual Conference of the Archaeological Association of Canada, 1993), 99–108.

2. M. N. Cohen and G. Armelagos, *Paleopathology at the Origins of Agriculture* (Gainesville: University Press of Florida, 1984).

3. F. Galton, "The First Steps towards the Domestication of Animals," *Transactions of the Ethnological Society of London*, 3 (1865): 122–138, 137.

4. Galton, "First Steps."
5. C. Darwin, *The Variation of Animals and Plants under Domestication* (London: John Murray, 1868), 36, 34.
6. See discussion in D. Rindos, *The Origins of Agriculture: An Evolutionary Perspective* (Sydney: Academic Press, 1984), 5–6.
7. The 2012 Harris Poll cited 95 percent of pet owners in the United States as considering their pets to be family members.
8. P. Ucko and G. W. Dimbleby, eds., *The Domestication and Exploitation of Plants and Animals* (Chicago: Aldine, 1969), xvi.
9. A. Sherratt, "Plough and Pastoralism: Aspects of the Secondary Products Revolution," in I. Hodder, G. Isaac, and N. Hammond, eds., *Pattern of the Past: Studies in Honour of David Clarke* (Cambridge: Cambridge University Press, 1981), 261–305; A. Sherratt, "The Secondary Exploitation of Animals in the Old World," *World Archaeology* 15, no. 1 (1983): 90–104.
10. P. Shipman, "And the Last Shall Be First," in H. Greenfield, ed., *Animal Secondary Products* (Oxford: Oxbow Books, 2014), 40–54; S. Bököyi, "Archaeological Problems and Methods of Recognizing Animal Domestication," in Ucko and Dimbleby, *Domestication and Exploitation,* 219.
11. M. A. Zeder, "Core Questions in Domestication Research," *Proceedings of the National Academy of Sciences* 112, no. 11 (2015): 191–198.
12. G. Larson and D. Fuller, "The Evolution of Animal Domestication," *Annual Review of Ecology, Evolution, and Systematics* 45: 115–136.
13. S. Bökönyi, "Development of Early Stock Rearing in the Near East," *Nature* 264 (1976): 19–23; S. Crockford, *Rhythms of Life: Thyroid Hormone and the Origin of Species* (Victoria, BC: Trafford Publishing, 2006).
14. R. Losey, T. Nomokonova, D. V. Arzyutov, et al., "Domestication as Enskilment: Harnessing Reindeer in Arctic Siberia," *Journal of Archaeological Method and Theory* 28 (2021): 197–231, 198.
15. R. J. Losey, V. I. Bazaliiskii, S. Garvie-Lok, et al., "Canids as Persons: Early Neolithic Dog and Wolf Burials, Cis-Baikal, Siberia," *Journal of Anthropological Archaeology* 30 (2011): 174–189.
16. M. Germonpré, M. Lázničková-Galetová, M. V. Sablin, and H. Bocherens, "Self-domestication or Human Control? The Upper Palaeolithic Domestication of the Wolf in Hybrid Communities," in C. Stepanoff and J.-D. Vigne, eds., *Biosocial Approaches to Domestication and Other Trans-species Relationships* (London: Routledge, 2018), 39–64.

17. J. Clutton-Brock, *A Natural History of Domesticated Animals* (Cambridge: Cambridge University Press, 1999).

18. D. F. Morey, "In Search of Paleolithic Dogs: A Quest with Mixed Results," *Journal of Archaeological Science* 52 (2014): 300–307; D. F. Morey and R. Jeger, "Paleolithic Dogs: Why Sustained Domestication Then?" *Journal of Archaeological Science* 3 (2015): 420–428.

6. Where Did the First Dog Come From?

1. R. Wayne, "Molecular Evolution of the Dog Family," *Trends in Genetics* 9, no. 6 (1993): 218–224.

2. S. Olsen, *Origins of the Domestic Dog: The Fossil Record* (Tucson: University of Arizona Press, 1985).

3. L. Janssens, A. Perri, P. Crombe, et al., "An Evaluation of Classical Morphologic and Morphometric Parameters Reported to Distinguish Wolves and Dogs," *Journal of Archaeological Science: Reports* 23 (2019): 501–533, 531, 533; P. Ciucci, V. Lucchini, L. Boitani, and E. Randi, "Dewclaws in Wolves as Evidence of Admixed Ancestry with Dogs," *Canadian Journal of Zoology* 81, no. 12 (2003): 2077–2081.

4. C. Vilà, P. Savolainen, J. E. Maldonado, et al., "Multiple and Ancient Origins of the Domestic Dog," *Science* 298 (1997): 1613–1616.

5. P. Savolainen, Y-P. Zhang, Luo Jing, et al., "Genetic Evidence for an East Asian Origin of Domestic Dogs," *Science* 298 (2002): 1610–1613.

6. S. Davis and F. Valla, "Evidence for Domestication of the Dog 12,000 Years Ago in the Natufian of Israel," *Nature* 276 (1978): 608–610.

7. Davis and Valla, "Evidence for Domestication of the Dog," 610.

8. K. Lord, "A Comparison of the Sensory Development of Wolves (*Canis lupus lupus*) and Dogs (*Canis lupus familiaris*)," *Ethology* 119, no. 2 (2013): 110–120.

9. M. Germonpré, M. V. Sablin, R. E. Stevens, et al., "Fossil Dogs and Wolves from Palaeolithic Sites in Belgium, the Ukraine and Russia: Osteometry, Ancient DNA and Stable Isotopes," *Journal of Archaeological Science* 36, no. 2 (2009): 473–490; M. Germonpré, M. Lázničková-Galetová, and M. Sablin, "Palaeolithic Dog Skulls at the Gravettian Předmostí Site, the Czech Republic," *Journal of Archaeological Science* 39, no. 1 (2012): 184–202; N. Ovodov, S. Crockford, Y. Kuzmin, et al., "A 33,000-Year-Old Incipient Dog from the Altai Mountains of Siberia: Evidence of the Earliest Domestication Disrupted by the Last Glacial Maximum," *PLoS ONE* 6, no. 7 (201): 22821.

10. O. Thalmann, B. Shapiro, P. Cui, et al., "Complete Mitochondrial Genomes of Ancient Canids Suggest a European Origin of Domestic Dogs," *Science* 342 (2013): 871.

7. Interwoven Stories

1. P. Shipman, *The Invaders: How Humans and Their Dogs Drove Neanderthals to Extinction* (Cambridge, MA: Belknap Press of Harvard University Press, 2015).

2. R. Losey, T. Komokonova, L. Fleming, et al., "Buried, Eaten, Sacrificed: Archaeological Dog Remains from Trans-Baikal," *Archaeological Research in Asia* 16 (2018): 58–65.

8. The Missing Dogs

1. P. Brown, T. Sutikna, M. J. Morwood, et al., "A New Small-Bodied Hominin from the Late Pleistocene of Flores, Indonesia," *Nature* 431 (2004): 1055–1061; F. Détroit, A. Mijares, J. Corny, et al., "A New Species of *Homo* from the Late Pleistocene of the Philippines," *Nature* 568 (2019): 181–186.

2. G. Hamm, P. Mitchell, L. Arnold, et al., "Cultural Innovation and Megafauna Interaction in the Early Settlement of Arid Australia," *Nature* 539 (2016): 280–283; S. Kealy, J. Louys, and S. O'Connor, "Islands under the Sea: A Review of Early Modern Human Dispersal Routes and Migration Hypotheses through Wallacea," *Journal of Island and Coastal Archaeology* 11 (2016): 364–384.

3. W. Noble and I. Davidson, "The Evolutionary Emergence of Modern Human Behavior," *Man* 26 (1991): 223–253; I. Davidson and W. Noble, "Why the First Colonisation of the Australian Region Is the Earliest Evidence of Modern Human Behaviour," *Archaeology in Oceania* 27 (1992): 135–142; I. Davison, "The Colonization of Australia and Its Adjacent Islands and the Evolution of Modern Cognition," *Current Anthropology* 51 (2010): s177–s189.

4. C. Marean, M. Bar-Matthews, J. Bernatchez, et al., "Early Human Use of Marine Resources and Pigment in South Africa during the Middle Pleistocene," *Nature* 449 (2007): 905–908.

5. C. Clarkson, Z. Jacobs, B. Marwick, et al., "Human Occupation of Northern Australia by 65,000 Years Ago," *Nature* 547 (2017): 306–325.

6. C. Marean, "How *Homo sapiens* Became the Ultimate Invasive Species," *Scientific American* 313, no. 2 (2015): 31–39; C. Marean, "The Origins

and Significance of Coastal Resource Use in Africa and Western Eurasia," Journal of Human Evolution 77 (2014): 17–40.

7. S. O'Connor, "New Evidence from East Timor Contributes to Our Understanding of Earliest Modern Human Colonisation East of the Sunda Shelf," *Antiquity* 81 (2007): 523–535; S. O'Connor, M. Spriggs, and P. Veth, "Excavation at Lene Hara Establishes Occupation in East Timor at Least 30,000–35,000 Years On: Results of Recent Fieldwork," *Antiquity* 76 (2002): 45–49; S. O'Connor and P. Veth, "Early Holocene Shell Fish Hooks from Lene Hara Cave, East Timor, Establish That Complex Fishing Technology Was in Use in Island South East Asia Five Thousand Years before Austronesian Settlement," *Antiquity* 79 (2005): 1–8.

8. S. O'Connor, R. Ono, and C. Clarkson, "Pelagic Fishing at 42,000 Years before the Present and the Maritime Skills of Modern Humans," *Science* 334, no. 6059 (2011): 1117–1121.

9. P. Veth, I. Ward, and S. O'Connor, "Coastal Feasts: A Pleistocene Antiquity for Resource Abundance in the Maritime Deserts of North West Australia?" *Journal of Island and Coastal Archaeology* 12 (2017): 8–23; P. Veth, K. Ditchfield, and F. Hook, "Maritime Deserts of the Australian Northwest," *Australian Archaeology* 79 (2014): 156–166; M. A. Bird, D. O'Grady, and S. Ulm, "Humans, Water, and the Colonization of Australia," *Proceedings of the National Academy of Sciences* 13, no. 41 (2016): 11477–11482, 11477.

10. J. Balme, "Of Boats and String: The Maritime Colonisation of Australia," *Quaternary International* 285 (2013): 68–75; J. Smith, "Did Early Hominids Cross Sea Gaps on Natural Rafts?" in I. Metcalfe, J. M. B. Smith, M. Morwood, and I. Davidson, eds., *Faunal and Floral Migration and Evolution in SE Asia-Australia* (Lisse, Netherlands: Swets & Zeitlinger, 2001), 409–416.

9. Adaptations

1. M. A. Bird, D. O'Grady, and S. Ulm, "Humans, Water, and the Colonization of Australia," *Proceedings of the National Academy of Sciences* 13, no. 41 (2016): 11477–11482, 11477; M. Bird, S. C. Condie, S. O'Connor, et al., "Early Human Settlement of Sahul Was Not an Accident," *Nature Scientific Reports* 9 (2019): 8220.

2. G. J. Price, "Taxonomy and Palaeobiology of the Largest-Ever Marsupial, *Diprotodon* Owen, 1838 (Diprotodontidae, Marsupialia)," *Zoological Journal of the Linnean Society* 153, no. 2 (2008): 369–397.

3. M. Bird, C. Turney, L. Fifield, et al., "Radiocarbon Analysis of the Early Archaeological Site of Nauwalabila 1, Arnhem Land, Australia: Implications for Sample Suitability and Stratigraphic Integrity," *Quaternary Science Reviews* 21 (2002): 1061–1075; J. F. O'Connell and J. Allen, "The Restaurant at the End of the Universe: Modelling the Colonisation of Sahul," *Australian Archaeology* 74 (2012): 5–17.

4. A. Thorne, E. Grün, G. Mortimer, et al., "Australia's Oldest Human Remains: Age of the Lake Mungo 3 Skeleton," *Journal of Human Evolution* 36 (1999): 591–612; J. M. Bowler, H. Johnston, J. M. Olley, et al., "New Ages for Human Occupation and Climatic Change at Lake Mungo, Australia," *Nature* 421 (2003): 837–841.

5. J. Balme, D. Merrilees, and J. Porter, "Late Quaternary Mammal Remains Spanning about 30,000 Years from Excavations in Devil's Lair, Western Australia," *Journal of the Royal Society of Western Australia* 60, no. 2 (1978): 33–65; C. Turney, M. I. Bird, L. K. Fifield, et al., "Early Human Occupation at Devil's Lair, Southwestern Australia 50,000 Years Ago," *Quaternary Research* 55 (2001): 3–13.

6. G. Hamm, P. Mitchell, L. Arnold, et al., "Cultural Innovation and Megafauna Interaction in the Early Settlement of Arid Australia," *Nature* 539 (2016): 280–283.

7. P. Hiscock, S. O'Connor, J. Balme, and T. Maloney, "World's Earliest Ground-Edge Axe Production Coincides with Human Colonisation of Australia," *Australian Archaeology* 82, no. 1 (2016): 2–11; M. Langley, S. O'Connor, and K. Aplin, "A >46,000-year-old Kangaroo Bone Implement from Carpenter's Gap 1 (Kimberley, Northwest Australia)," *Quaternary Science Reviews* 154 (2016): 199–213; S. O'Connor, "Carpenter's Gap Rock Shelter 1: 40,000 Years of Aboriginal Occupation in the Napier Ranges, Kimberley, WA," *Australian Archaeology* 40 (2014): 58–60.

8. C. Shipton, S. O'Connor, S. Kealy, et al., "Early Ground Axe Technology in Wallacea: The First Excavations on Obi Island," *PLoS ONE* 15, no. 8 (2020): e0236719.

9. Langley et al., "A >46,000-Year-Old Kangaroo Bone Implement."

10. Hiscock et al., "World's Earliest Ground-Edge Axe Production," 9.

11. I. Davidson, "Peopling the Last New Worlds: The First Colonisation of Sahul and the Americas," *Quaternary International* 285 (2013): 1–29.

12. S. Kealy, J. Louys, and S. O'Connor, "Least-Cost Pathway Models Indicate Northern Human Dispersal from Sunda to Sahul," *Journal of Human Evolution* 125 (2018): 59–70.

13. R. Gillespie, "Dating the First Australians," *Radiocarbon* 44, no. 20 (2020): 455–472.

14. M. Williams, N. A. Spooner, K. McDonnell, and J. F. O'Connell, "Identifying Disturbance in Archaeological Sites in Tropical Northern Australia: Implications for Previously Proposed 65,000-Year Continental Occupation Date," *Geoarchaeology* 36, no. 1 (2021): 92–108, 105; O'Connell and. Allen, "Restaurant at the End of the Universe"; J. O'Connell, J. Allen, M. Williams, et al., "When Did *Homo sapiens* First Reach Southeast Asia and Sahul?" *Proceedings of the National Academy of Sciences* 115, no. 34 (2018): 8482–8490.

10. Surviving in New Ecosystems

1. S. J. Wroe, "Australian Marsupial Carnivores: Recent Advances in Palaeontology," in M. Jones, C. Dickman, and M. Archer, eds., *Predators with Pouches: The Biology of Carnivorous Marsupials* (Collingwood, Victoria: CSIRO Publishing, 2003), 102–123; D. A. Rovinsky, A. R. Evans, D. G. Martin, and J. W. Adams, "Did the Thylacine Violate the Costs of Carnivory? Body Mass and Sexual Dimorphism of an Iconic Australian Marsupial," *Proceedings of the Royal Society B: Biological Sciences* 287 (2020): 20201537.

2. A. Gonzalez, G. Clark, S. O'Connor, and L. Matisoo-Smith, "A 3000 Year Old Dog Burial in Timor-Leste," *Australian Archaeology* 76, no. 1 (2013): 13–20.

3. R. Paddle, *The Last Tasmanian Tiger* (Cambridge: Cambridge University Press, 2003); D. Owen, *Thylacine: The Tragic Tale of the Tasmanian Tiger* (Baltimore: Johns Hopkins University Press, 2004).

4. K. Akerman and T. Willing, "An Ancient Rock Painting of a Marsupial Lion, *Thylacoleo carnifex*, from the Kimberley, Western Australia," *Antiquity* 83 (2009): 319; K. Akerman, "Interaction between Humans and Megafauna Depicted in Australian Rock Art?" *Antiquity,* Project Gallery, vol. 83, no. 322 (2009).

5. A. Goswami, N. Milne, and S. Wroe, "Biting through Constraints: Cranial Morphology, Disparity and Convergence across Living and Fossil Carnivorous Mammals," *Proceedings of the Royal Society B: Biological Sciences* 278 (2011): 1831–1839.

6. R. T. Wells and A. Camens, "New Skeletal Material Sheds Light on the Palaeobiology of the Pleistocene Marsupial Carnivore, *Thylacoleo carnifex,*" *PLoS One E* 13, no. 12 (2018): 0208020; D. Horton

and R. Wright, "Cuts on Lancefield Bones: Carnivorous *Thylacoleo*, Not Humans, the Cause," *Archaeology in Oceania* 16, no. 2 (1981): 73–80.

7. S. Wroe, C. Argot, and C. Dickman, "On the Scarcity of Big Fierce Carnivores and Primacy of Isolation and Area: Tracking Large Mammalian Predator Diversity of Two Isolated Continents," *Proceedings of the Royal Society B: Biological Sciences* 217 (2002): 1203–1211.

11. Why Has the Australian Story Been Overlooked for So Long?

1. B. Griffiths, "'The Dawn' of Australian Archaeology: John Mulvaney at Fromm's Landing," *Journal of Pacific Archaeology* 8, no. 1 (2017): 100–111.

2. R. Hughes, *The Fatal Shore* (New York: Knopf, 1986), 84.

3. S. M. van Holst Pellekaan, "Genetic Research: What Does This Mean for Indigenous Australian Communities?" *Journal of Australian Aboriginal Studies* 1–2 (2000): 65–75.

4. R. Pullein, "The Tasmanians and Their Stone Culture," *Australasian Association for the Advancement of Science* 19 (1928): 294–314; I. Davidson, "A Lecture by the Returning Chair of Australian Studies, Harvard University 2008–2009: Australian Archaeology as a Historical Science," *Journal of Australian Studies* 34, no. 3 (2010): 377–398, 388; Griffiths, "'Dawn' of Australian Archaeology."

5. G. Hamm, P. Mitchell, L. Arnold, et al., "Cultural Innovation and Megafauna Interaction in the Early Settlement of Arid Australia," *Nature* 539 (2016): 280–283.

12. The Importance of Dingoes

1. M. Fillios and P. Taçon, "Who Let The Dogs In? A Review of the Recent Genetic Evidence for the Introduction of the Dingo to Australia and Implications for the Movement of People," *Journal of Archaeological Science: Reports* 7 (2016): 782–792; A. R. Boyko, R. H. Boyko, C. M. Boyko, et al., "Complex Population Structure in African Village Dogs and Its Implications for Inferring Dog Domestication History," *Proceedings of the National Academy of Sciences* 106 (2009): 13903–13908; L. Shannon, R. Boyko, M. Castelhanoc, et al., "Genetic Structure in Village Dogs Reveals a Central Asian Domestication Origin,"

Proceedings of the National Academy of Sciences 112, no. 44 (2015): 13639–13644.

2. A. Ardalan, M. Oskarsson, C. Natanaelsson, et al., "Narrow Genetic Basis for the Australian Dingo Confirmed through Analysis of Paternal Ancestry," *Genetica* 140 (2012): 65–73; P. Savolainen, T. Leitner, A. Wilton, et al., "A Detailed Picture of the Origin of the Australian Dingo, Obtained from the Study of Mitochondrial DNA," *Proceedings of the National Academy of Sciences* 101 (2004): 12387–12390; K. Cairns and A. Wilton, "New Insights on the History of Canids in Oceania Based on Mitochondrial and Nuclear Data," *Genetica* 144 (2016): 553–565.

3. J. McIntyre, L. Wolf, B. Sacks, et al., "A Population of Free-Living Highland Wild Dogs in Indonesian Papua," *Australian Mammalogy* 42, no. 2 (2019): 160–166.

4. S. Surbakti, H. Parker, J. McIntyre, et al., "New Guinea Highland Wild Dogs Are the Original New Guinea Singing Dogs," *Proceedings of the National Academy of Sciences* 117 (2020): 24369–24376.

5. B. Gammage, *The Biggest Estate on Earth: How Aborigines Made Australia* (Sydney: Allen & Unwin, 2011).

6. J. Boyce, "Canine Revolution: The Social and Environmental Impact of the Introduction of the Dog to Tasmania," *Environmental History* 11, no. 1 (2006): 102–129. Much of the discussion that follows is after J. Boyce, *Van Dieman's Land,* 2nd ed. (Melbourne: Black, 2010), and archival sources cited therein.

7. E. Matisoo-Smith, "The Human Colonisation of Polynesia. A Novel Approach: Genetic Analyses of the Polynesian Rat (*Rattus exulans*)," *Journal of the Polynesian Society* 103 (1994): 75–87.

8. L. Corbett, "The Conservation Status of the Dingo *Canis lupus dingo* in Australia, with Particular Reference to New South Wales: Threats to Pure Dingoes and Potential Solutions," in C. Dickman and D. Lunney, eds., *The Dingo Dilemma: A Symposium on the Dingo* (Sydney: Royal Zoological Society of New South Wales, 2001), 10–19; L. Shannon, R. Boyko, M. Castelhanoc, et al., "Genetic Structure in Village Dogs Reveals a Central Asian Domestication Origin," *Proceedings of the National Academy of Sciences* 112, no. 44 (2015): 13639–13644.

9. S. Zhang, G-D. Wang, M. Pengcheng, et al., "Genomic Regions under Selection in the Feralization of the Dingoes," *Nature Communications* 11 (2020): 671.

10. B. Sacks, A. Brown, D. Stephens, et al., "Y Chromosome Analysis of Dingoes and Southeast Asian Village Dogs Suggests a Neolithic Con-

tinental Expansion from Southeast Asia Followed by Multiple Austronesian Dispersals," *Molecular Biology and Evolution* 30, no. 5 (2013): 1103–1118; K. Cairns, S. Brown, B. Sacks, and J. Ballard, "Conservation Implications for Dingoes from the Maternal and Paternal Genome: Multiple Populations, Dog Introgression, and Demography," *Ecology and Evolution* 7 (2017): 9787–9807.

11. See review in P. Shipman, "What Does the Dingo Say about Dog Domestication?" *Anatomical Record* 304 (2021): 19–30.

13. How Invasion Works

1. J. Balme, S. O'Connor, and S. Fallon, "New Dates on Dingo Bones from Madura Cave Provide Oldest Firm Evidence for Arrival of the Species in Australia," *Nature Scientific Reports* 8 (2018): 9933–9939; K. Gollan, "The Australian Dingo: In the Shadow of Man," in M. Archer and G. Clayton, eds., *Vertebrate Zoogeography and Evolution in Australasia* (Perth: Hesperian Press, 1984), 921–927; G. Saunders, B. Coman, J. Kinnear, and M. Braysher, *Managing Vertebrate Pests: Foxes* (Canberra: Australian Government Publishing Service, 1995); I. Abbott, "The Spread of the Cat, *Felis catus,* in Australia: Re-examination of the Current Conceptual Model with Additional Animals," *Conservation Science Western Australia* 7, no. 1 (2008): 1–17.

2. A. Elledge, L. Allen, B-L. Carlsson, et al., "An Evaluation of Genetic Analyses, Skull Morphology and Visual Appearance for Assessing Dingo Purity: Implications for Dingo Conservation," *Wildlife Research* 35 (2008): 812–820; W. Parr, L. Wilson, S. Wroe, et al., "Cranial Shape and the Modularity of Hybridization in Dingoes and Dogs: Hybridization Does Not Spell the End for Native Morphology," *Evolutionary Biology* 43, no. 2 (2016): 171–187; A. Wilton, "DNA Methods of Assessing Australian Dingo Purity," in C. R. Dickman and D. Lunney, eds., *A Symposium on the Australian Dingo* (Sydney: Royal Zoological Society of New South Wales, 2017), 49–55; A. N. Wilton, D. J. Steward, and K. Zafaris, "Microsatellite Variation in the Australian Dingo," *Journal of Heredity* 90 (1999): 108–111; A. Ardalan, M. Oskarsson, C. Natanaelsson, et al., "Narrow Genetic Basis for the Australian Dingo Confirmed through Analysis of Paternal Ancestry,"*Genetica* 140 (2012): 65–73; K. Cairns and A. Wilton, "New Insights on the History of Canids in Oceania Based on Mitochondrial and Nuclear Data," *Genetica* 144 (2016): 553–565.

3. S. Surbakti, H. Parker, J. McIntyre, et al., "New Guinea Highland Wild Dogs Are the Original New Guinea Singing Dogs," *Proceedings of the National Academy of Sciences* 117, no. 39 (2020): 24369–24376; Ardalan et al., "Narrow Genetic Basis for the Australian Dingo."

4. D. Rose, *Dingo Makes Us Human* (New York: Cambridge University Press, 1992), 176–177.

5. B. Allen, "Do Desert Dingoes Drink Daily? Visitation Rates at Remote Waterpoints in the Strzelecki Desert," *Australian Mammalogy* 34, no. 2 (2011): 251–256; C. Hicks, "The Australian Aboriginal: A Study in Comparative Physiology," *Schweizerische Medizinische Wochenschrift* 71, no. 12 (1941): 385–388.

6. J. Balme and S. O'Connor, "Dingoes and Aboriginal Social Organization in Holocene Australia," *Journal of Archaeological Science: Reports* 7 (2016): 775–781.

7. R. A. Breckwold, *A Very Elegant Animal: The Dingo* (North Ryde, NSW: Angus & Robertson, 1988); Rose, *Dingo Makes Us Human*, 176–177.

8. The duality of the dingoes' role in Aboriginal peoples' thought is extensively discussed in M. Parker, "Bringing the Dingo Home: Discursive Representations of the Dingo by Aboriginal, Colonial, and Contemporary Australians" (B.A. honors thesis, University of Tasmania, 2006); F. Clark and I. Cahir, "The Historic Importance of the Dingo in Aboriginal Society in Victoria (Australia): A Reconsideration of the Archival Record," *Anthrozoös* 26, no. 2 (2013): 185–198, 193.

9. R. Gunn, R. Whear, and L. Douglas, "A Dingo Burial from the Arnhem Land Plateau," *Australian Archaeology* 71 (2010): 11–16.

10. Gunn et al., "A Dingo Burial," 12 (remark by Jacob Nayinggul, Kunwinggu elder, pers. comm. to Gunn, 1992).

11. G. Chaloupka, *Journey in Time: The World's Longest Continuing Art Tradition* (Chatswood, NSW: Reed, 1993); R. Gunn, R. Whear, and L. Douglas, "A Second Recent Canine Burial from the Arnhem Land Plateau," *Australian Archaeology* 71 (2010): 103–105; E. Kolig, "Aboriginal Man's Best Foe," *Mankind* 9, no. 2 (1973): 122–123; B. Griffiths, "'The Dawn' of Australian Archaeology: John Mulvaney at Fromm's Landing," *Journal of Pacific Archaeology* 8, no. 1 (2017): 100–111.

12. Parker, "Bringing the Dingo Home"; M. Fillios and P. Taçon, "Who Let the Dogs In? A Review of the Recent Genetic Evidence for the Introduction of the Dingo to Australia and Implications for the Movement of People," *Journal of Archaeological Science: Reports* 7 (2016): 782–792.

13. Breckwold, *A Very Elegant Animal*.

14. See, e.g., P. Veth, N. Stern, J. McDonald, et al., "The Role of Information Exchange in the Colonization of Sahul," in R. Whallon, W. Lovis, and R. Hitchcock, eds., *Information and Its Role in Hunter-Gatherer Bands* (Los Angeles: Cotsen Institute of Archaeology Press, 2011), 203–220.

15. J. Balme, I. Davidson, J. McDonald, et al., "Symbolic Behaviour and the Peopling of the Southern Arc Route to Australia," *Quaternary International* 202 (2009): 59–68.

16. P. Nunn and N. Reid, "Aboriginal Memories of Inundation of the Australian Coast Dating from More than 7000 Years Ago," *Australian Geographer* 47, no. 1 (2015): 1–37.

17. B. Pascoe, *Dark Emu: Aboriginal Australia and the Birth of Agriculture* (London: Scribe Publications, 2018).

18. P. Savolainen, P. Milheim, and P. Thompson, "Relative Antiquity of Human Occupation and Extinct Fauna at Madura Cave, Southeastern Western Australia," *Mankind* 10, no. 3 (1976): 175–180; J. Balme, S. O'Connor, and S. Fallon, "New Dates on Dingo Bones from Madura Cave Provide Oldest Firm Evidence for Arrival of the Species in Australia," *Scientific Reports* 8 (2018): 9933–9939.

19. M. Letnic, M. Fillios, and M. S. Crowther, "The Arrival and Impacts of the Dingo," in A. Glen and C. Dickman, eds., *Carnivores of Australia: Past, Present, and Future* (Clayton, Victoria: CSIRO Publishing, 2014), 53–68.

20. M. Letnic, M. Fillios, and M. S. Crowther, "Could Direct Killing by Larger Dingoes Have Caused the Extinction of the Thylacine from Mainland Australia?" *PLoS One* 7, no. 1 (2012): 34877–34882; M. Fillios, M. Crowther, and M. Letnic, "The Impact of the Dingo on the Thylacine in Holocene Australia," *World Archaeology* 44, no. 1 (2018): 118–134.

21. L. Koungolous and M. Fillios, "Hunting Dogs Down Under? On the Aboriginal Use of Tame Dingoes in Dietary Game Acquisition and Its Relevance to Australian Prehistory," *Journal of Anthropological Archaeology* 58 (2020): 101146.

22. A. Pope, C. Grigg, S. Cairns, et al., "Trends in the Numbers of Red Kangaroos and Emus on Either Side of the South Australian Dingo Fence: Evidence for Predator Regulation?" *Wildlife Research* 27 (2000): 269–276; A. Glen, C. R. Dickman, R. E. Soulé, and B. Mackey, "Evaluating the Role of the Dingo as a Trophic Regulator in Australian Ecosystems," *Austral Ecology* 32, no. 5 (2007): 492–501.

23. C. Johnson, J. Isaac, and D. Fisher, "Rarity of a Top Predator Triggers Continent-Wide Collapse of Mammal Prey: Dingoes and Marsupials in Australia," *Proceedings of the Royal Society B: Biological Sciences* 274 (2007): 341–346; M. Letnic, E. Ritchie, and C. Dickman, "Top Predators as Biodiversity Regulators: The Dingo *Canis lupus dingo* as a Case Study," *Biological Reviews* 87 (2012): 390–413.

14. A Different Story

1. V. Pitulko, P. Nikolsky, E. Girya, et al., "The Yana RHS Site: Humans in the Arctic before the Last Glacial Maximum," *Science* 303, no. 5654 (2004): 52–56.

2. A. Stone, "Human Lineages in the Far North," *Nature* 570 (2019): 170–172.

3. D. Meltzer, D. Grayson, G. Ardilla, et al., "On the Pleistocene Antiquity of Monte Verde, Southern Chile," *American Antiquity* 62, no. 4 (1997): 659–663, 662.

4. T. Dillehay, *Monte Verde: A Late Pleistocene Site in Chile*, vol. 2 (Washington, DC: Smithsonian Institution Press, 1997).

5. J. Erlandson, "After Clovis-First Collapsed: Reimagining the Peopling of the Americas," in K. Graf, C. Ketron, and M. Waters, eds., *PaleoAmerican Odyssey* (College Station: Texas A&M University Press, Center for the Study of the First Americans, 2013), 127–132, 127.

6. R. Tamm, M. Reidla, M. Metspalu, et al., "Beringian Standstill and Spread of Native American Founders," *PLoS One* 2, no. 9 (2007): e829; B. Potter, J. Irish, J. Reuther, J., et al., "Terminal Pleistocene Child Cremation and Residential Structure from Eastern Beringia," *Science* 331 (2011): 1058–1062.

7. J. V. Moreno-Mayar, B. Potter, V. Lasse Vinner, et al., "Terminal Pleistocene Alaskan Genome Reveals First Founding Population of Native Americans," *Nature* 553 (2018): 203–208; J. Tackney, B. Potter, J. Raff, et al., "Two Contemporaneous Mitogenomes from Terminal Pleistocene Burials in Eastern Beringia," *Proceedings of the National Academy of Sciences* 112, no. 45 (2015): 13833–13838.

8. A. Bergström, L. Frantz, R. Schmidt, et al., "Origins and Genetic Legacy of Prehistoric Dogs," *Science* 370 (2020): 557–564.

9. G. Perry, N. Dominy, K. Claw, et al., "Diet and the Evolution of Human Amylase Gene Copy Number Variation," *Nature Genetics* 39 (2007): 1256–1260.

15. Heading North

1. The discussion following relies heavily on information reported in R. V. Losey, S. Garvie-Lok, M. Germonpré, et al., "Canids as Persons: Early Neolithic Dog and Wolf Burials, Cis-Baikal, Siberia," *Journal of Anthropological Archaeology* 30 (2011): 174–189; R. Losey, S. Garvie-Lok, J. A. Leonard, et al., "Burying Dogs in Ancient Cis-Baikal, Siberia: Temporal Trends and Relationships with Human Diet and Subsistence Practices," *PLoS One* 8, no. 5 (2013): e63740; R. Losey, T. Nomokonova, L. Fleming, et al., "Buried, Eaten, Sacrificed: Archaeological Dog Remains from Trans-Baikal, Siberia," *Archaeological Research in Asia* 16 (2018): 58–65.

2. V. I. Bazaliiskiy and N. A. Savelyev, "The Wolf of Baikal: the 'Lokomotiv' Early Neolithic Cemetery in Siberia (Russia)," *Antiquity* 77 (2003): 20–30.

3. I. Paulsen, "The Preservation of Animal-Bones in the Hunting Rites of Some North-Eurasian people," in V. Dioszegi, ed., *Popular Beliefs and Folklore Traditions in Siberia* (The Hague: Mouton and Co., 1968), 448–451; Bazaliiskiy and Savelyev, "Wolf of Baikal"; A. W. Weber, "The Neolithic and Early Bronze Age of the Lake Baikal Region, Siberia: A Review of Recent Research," *Journal of World Prehistory* 9, no. 1 (1995): 99–165; A. Perri, "A Typology of Dog Deposition in Archaeological Contexts," in P. Rowley-Conwy, P. Halstead, and D. Serjeantson, eds., *Economic Zooarchaeology: Studies in Hunting, Herding and Early Agriculture* (Oxford: Oxbow Books, 2017), chap. 11.

4. V. Pitulko, V. Ivanova, A. Kasparov, and E. Pavlova, "Reconstructing Prey Selection, Hunting Strategy and Seasonality of the Early Holocene Frozen Site in the Siberian High Arctic: A Case Study on the Zhokhov Site Faunal Remains, De Long Islands," *Environmental Archaeology* 20, no. 2 (2015): 120–157.

5. A. Bergström, L. Frantz, R. Schmidt, et al., "Genetics and Origin of Prehistoric Dogs," *Science* 370, no. 6516 (2020): 557–564.

6. L. Loog, O. Thalmann, M.-H. Sinding, et al., "Modern Wolves Trace Their Origin to a Late Pleistocene Expansion from Beringia," *Molecular Ecology* 29, no. 9 (2020): 1596–1610.

7. C. Ameen, T. Feuerborn, S. Brown, et al., "Specialized Sledge Dogs Accompanied Inuit Dispersal across the North American Arctic," *Proceedings of the Royal Society B: Biological Sciences* 286, no. 1916 (2019), https://doi.org/10.1098/rspb.2019.1929.

8. M. Ní Leathlobhair, A. Perri, E. Irving-Pease, et al., "The Evolutionary History of Dogs in the Americas," *Science* 361 (2018): 81–85.

9. J. Leonard, R. Wayne, J. Wheeler, et al., "Ancient DNA Evidence for Old World Origin of New World Dogs," *Science* 289 (2002): 1613–1616.

10. A. Perri, T. Feuerborn, L. Frantz, et al., "Dog Domestication and the Dual Dispersal of People and Dogs into the Americas," *Proceedings of the National Academy of Sciences* 118 (2021): 201003118.

16. To the End of the Earth

1. F. Perini, C. Russo, and C. Schrago, "The Evolution of South American Endemic Canids: A History of Rapid Diversification and Morphological Parallelism," *Journal of Evolutionary Biology* 23 (2010): 311–332; P. Stahl, "Early Dogs and Endemic South American Canids of the Spanish Main," *Journal of Anthropological Research* 69 (2013): 515–533.

2. D. Kleiman, "Social Behavior of the Maned Wolf (*Chrysocyon brachyurus*) and Bush Dog (*Speothos venaticus*): A Study in Contrast," *Journal of Mammalogy* 53, no. 4 (1972): 791–806; A. Berta, "*Cerdocyon thous*," *Mammalian Species*, no. 186 (1982): 1–4; B. de Mello Beisiegel and G. Zuercher, "*Speothos venaticus*," *Mammalian Species*, no. 783 (2005): 1–6.

3. G. Slater, O. Thalmann, J. Leonard, et al., "Evolutionary History of the Falklands Wolf," *Current Biology* 19, no. 20 (2009): R937–R938.

4. C. Posth, N. Nakatsuka, I. Lazaridis, et al., "Reconstructing the Deep Population History of Central and South America," *Cell* 175 (2018): 1185–1197; M. Raghavan, M. Steinrücken, K. Harris, et al., "Genomic Evidence for the Pleistocene and Recent Population History of Native Americans," *Science* 349 (2015): 1185–1197; P. Skoglund, S. Mallick, M. Bortolini, et al., "Genetic Evidence for Two Founding Populations of the Americas," *Nature* 525 (2015): 104–108.

5. F. Vander Velden, "Narrating the First Dogs: Canine Agency in the First Contacts with Indigenous Peoples in the Brazilian Amazon," *Anthrozoös* 30, no. 4 (2017): 533–548.

6. Vander Velden, "Narrating the First Dogs."

7. J. Koster and K. Tankersley, "Heterogeneity of Hunting Ability and Nutritional Status among Domestic Dogs in Lowland Nicaragua," *Proceedings of the National Academy of Sciences* 109, no. 8 (2012):

E463–E470; J. Koster, "Hunting Dogs in the Lowland Neotropics," *Journal of Anthropological Research* 65 (2009): 575–610.

8. C. Wynne, *Dog Is Love: Why and How Your Dog Loves You* (New York: Houghton Mifflin Harcourt, 2019); B. vonHoldt, J. Pollinger, K. Lohmuelle, et al., "Genome-wide SNP and Haplotype Analyses Reveal a Rich History Underlying Dog Domestication," *Nature* 464 (2010): 898–903.

Acknowledgments

I could not have written this book without considerable help from many others. These include Cheryl Glenn, Jon Olson, Marian Copeland, Carol Phillips, Helga Vierich, Iain Davidson, Melanie Fillios, Sue O'Connor, Jane Balme, Vladimir Pitulko, Robert Losey, Mietje Germonpré, Bob Wayne, Blaire Van Valkenburgh, Bridgett vonHoldt, Chris Mason, Mike Waters, Tom Dillehay, Bradley Smith, Lyn Watson, Leigh Mullen, Jeffrey Mathison, Greg Retallack, Nina Jablonsky, George Chaplin, Mara Connolly Taft, and "Mack" McIntyre. I have probably forgotten a few steadfast friends and inspirations, but not by intent. I cannot begin to thank the many scholars whose brilliant works have meant a great deal to me. Thank you all.

Illustration Credits

p. 152   Photos by R. G. Gunn courtesy of the Jawoyn Association.

p. 162   Mike Letnic, Melanie Fillios, and Mathew S. Crowther, "Could Direct Killing by Larger Dingoes Have Caused the Extinction of the Thylacine from Mainland Australia?" *PLoS ONE* 7(5); e34877, fig. 1.

p. 168   Vladimir V. Pitulko, Elena Y. Pavlova, Pavel A. Nikolskiy, and Varvara V. Ivanova, "The Oldest Art of the Eurasian Arctic: Personal Ornaments and Symbolic Objects from Yana RHS, Arctic Siberia," *Antiquity* 86, no. 333 (2015): 642–659, fig. 9. Reprinted by permission of Cambridge University Press.

p. 171   Center for the Study of the First Americans, Texas A&M University.

p. 173   Courtesy of Tom D. Dillehay.

p. 183   Robert J. Losey, Vladimir I. Bazaliiskii, Sandra Garvie-Lok, Mietje Germonpré, Jennifer A. Leonard, Andrew L. Allen, Anne Katzenberg, and Mikhail Sablin, "Canids as Persons: Early Neolithic Dog and Wolf Burials, Cis-Baikal, Siberia," *Journal of Anthropological Archaeology* 30: 2 (2011): 174–189, fig. 4. Reprinted by permission of Elsevier.

p. 198   Imagebroker / Alamy Stock Photo.

p. 198   Azoor Wildlife Photo / Alamy Stock Photo.

p. 199   Avalon / Picture Nature / Alamy Stock Photo.

Index

hominins (*continued*)
Homo luzonensis, 89; Jōmon,
18, 75. *See also* Denisovans;
early modern humans;
Neanderthals
hormone levels, domestication
and, 60
horses, 163–164
Horton, David, 116
How the Dog Became the Dog
(Derr), 22
Hughes, Robert, 123
human migrations. *See* migration,
human
humans: dogs as proxies for, 2;
dogs' relationship with, 22,
24–26; living with, 30, 34;
treatment of dogs by, 55–56;
WBS in, 32–33; wolf-dogs'
selection of, 26–28; wolves'
relationship with, 22, 24–26
humans, early. *See* early modern
humans
humans, modern: Denisovan
genes in, 40–42; relationship
with Neanderthals, 38, 42
hunting, 59; in Australia,
134–135, 143, 163; commensal
evolution and, 62; dingoes and,
149–150, 161, 163; distance
weapons and, 27, 62, 187; dogs
and, 18–19, 55, 85, 112, 187,
202, 203, 204; domestication
and, 63; by early humans, 27, 28;
in Eurasia, 112; vs. farming,
51–52; by Native Americans,
202; in Nicaragua, 203; of
polar bears, 186–187; with
wolf-dogs, 18–19; wolves
and, 19, 62; by Zhokhov

people, 186–188. *See also*
cooperation/collaboration
hunting dogs, 85

Indigenous Australians. *See* Australians, First/Indigenous
Indonesia, 88, 89
intentionality, in domestication,
58
intimacy, physical, 26
introgression, 12
Inuit peoples, 190, 191, 192
Invaders, The (Shipman), 20, 84
invasion. *See* migration, human
island dwarfing, 88–89

jackals, 4, 84
Janssens, Luc, 69
Japan, Jōmon culture in, 18, 75.
See also Asia
Jeger, Rujana, 63
jobs, breeding dogs for, 189–190,
192
Jōmon culture, 18, 75

kangaroos, hunting of, 134–135
Kealy, Shimona, 45
Kesserloch dog, 73, 75
Khotorok grave, 183
Knopwood, Robert, 134, 135
knowledge: communicating,
79–81; of First Australians,
101, 119, 127, 128, 130, 132,
150–151, 155–156, 157–158,
159, 174, 206; sharing of, 127;
survival of, 158–159
Koster, Jeremy, 203
Koster dogs, 193
Koungoulos, Loukas, 163
Krause, Johannes, 36